The bending and stretching of plates

The bending and stretching of plates

Second edition

E. H. MANSFIELD

Formerly Chief Scientific Officer
Royal Aircraft Establishment
Farnborough

CAMBRIDGE UNIVERSITY PRESS

CAMBRIDGE
NEW YORK NEW ROCHELLE MELBOURNE SYDNEY

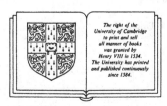

CAMBRIDGE UNIVERSITY PRESS
Cambridge, New York, Melbourne, Madrid, Cape Town, Singapore, São Paulo

Cambridge University Press
The Edinburgh Building, Cambridge CB2 2RU, UK

Published in the United States of America by Cambridge University Press, New York

www.cambridge.org
Information on this title: www.cambridge.org/9780521333047

First published 1964 by Pergamon Press
Second edition first published 1989
This digitally printed first paperback version 2005

A catalogue record for this publication is available from the British Library

Library of Congress Cataloguing in Publication data

Mansfield, Eric Harold. 1923–
The bending and stretching of plates/E.H. Mansfield. – 2nd ed.
p. cm.
Includes bibliographies and indexes.
ISBN 0-521-33304-0
1. Elastic plates and shells. I. Title.
QA935.M36 1989 88–29035
624.1′776–dc 19 CIP

ISBN-13 978-0-521-33304-7 hardback
ISBN-10 0-521-33304-0 hardback

ISBN-13 978-0-521-01816-6 paperback
ISBN-10 0-521-01816-1 paperback

CONTENTS

PREFACE

In the first edition of this book, I attempted to present a concise and unified introduction to elastic plate theory. Wherever possible, the approach was to give a clear physical picture of plate behaviour. The presentation was thus geared more towards engineers than towards mathematicians, particularly to structural engineers in aeronautical, civil and mechanical engineering and to structural research workers. These comments apply equally to this second edition. The main difference here is that I have included thermal stress effects, the behaviour of multi-layered composite plates and much additional material on plates in the large-deflexion régime. The objective throughout is to derive 'continuum' or analytical solutions rather than solutions based on numerical techniques such as finite elements which give little direct information on the significance of the structural design parameters; indeed, such solutions can become simply number-crunching exercises that mask the true physical behaviour.

E.H. Mansfield

PRINCIPAL NOTATION

a, b	typical plate dimensions
$\mathbf{a}, \mathbf{b}, \mathbf{d}$	defined by (1.96)
$\mathbf{A}, \mathbf{B}, \mathbf{D}$	defined by (1.93)
D	flexural rigidity $Et^3/\{12(1-\nu^2)\}$
E, G	Young modulus and shear modulus
$\mathbf{E}$	defined by (1.89)
k	foundation modulus
$\mathrm{L}_1, \mathrm{L}_2, \mathrm{L}_3$	differential operators defined after (1.98)
M_x, M_y, M_{xy}; $M_r, M_\theta, M_{r\theta}$	bending and twisting moments per unit length
$\mathbf{M}$	$(M_x, M_y, M_{xy})^{\mathrm{T}}$
$\mathscr{M}_a$	total moment about a generator
$\boldsymbol{n}$	normal to boundary
N_x, N_y, N_{xy}	direct and shear forces per unit length in plane of plate
$\mathbf{N}$	$(N_x, N_y, N_{xy})^{\mathrm{T}}$
P	point load
Q_x, Q_y; Q_r, Q_θ	transverse shear forces per unit length
q	transverse loading per unit area
q_n, q_{mn}	coefficients in Fourier expansions for q
r, θ, z	cylindrical coordinates, r, θ in plane of plate
s	distance along boundary
t	plate thickness
$\boldsymbol{t}$	time, or tangent to boundary
T	temperature, torque
U	strain energy
u, v, w	displacements in x, y, z directions
w_1, w_p	particular integrals
w_2, w_c	complementary functions
x, y, z	Cartesian coordinates, x, y in plane of plate
α	coefficient of thermal expansion, angle between generator and x-axis

$\varepsilon_x, \varepsilon_y, \varepsilon_{xy}$	direct and shear strains in plane $z = \text{const.}$
ε_T, κ_T	thermal strain and curvature, see (1.56) and (1.57)
$\boldsymbol{\varepsilon}$	$(\varepsilon_x, \varepsilon_y, \varepsilon_{xy})^{\mathrm{T}}$
η	distance along a generator
$\kappa_x, \kappa_y, \kappa_{xy}$	curvatures, $-(\partial^2 w/\partial x^2)$, etc.
κ	$(\kappa_x, \kappa_y, 2\kappa_{xy})^{\mathrm{T}}$
ν	Poisson ratio
Π	potential energy
ϱ	transverse edge support stiffness, or r/r_1
$\sigma_x, \sigma_y, \tau_{xy}$	direct and shear stresses in plane $z = \text{const.}$
$\boldsymbol{\sigma}$	$(\sigma_x, \sigma_y, \sigma_{xy})^{\mathrm{T}}$
φ	complex potential function, or $-\mathrm{d}w/\mathrm{d}r$
Φ	force function
χ	rotational edge support stiffness, or complex potential function
ψ	angle between tangent to boundary and x-axis

$$\nabla^2 f = \frac{\partial^2 f}{\partial x^2} + \frac{\partial^2 f}{\partial y^2}$$

$$\begin{aligned}\diamond^4(f, g) &= \frac{\partial^2 f}{\partial x^2}\frac{\partial^2 g}{\partial y^2} - 2\frac{\partial^2 f}{\partial x \partial y}\frac{\partial^2 g}{\partial x \partial y} + \frac{\partial^2 f}{\partial y^2}\frac{\partial^2 g}{\partial x^2} \\ &\equiv \tfrac{1}{2}\{(\nabla^2 f)(\nabla^2 g) + \nabla^2(f\nabla^2 g + g\nabla^2 f)\} \\ &\quad - \tfrac{1}{4}\{\nabla^4(fg) + f\nabla^4 g + g\nabla^4 f\}\end{aligned}$$

I

SMALL-DEFLEXION THEORY

1

Derivation of the basic equations

All structures are three-dimensional, and the exact analysis of stresses in them presents formidable difficulties. However, such precision is seldom needed, nor indeed justified, for the magnitude and distribution of the applied loading and the strength and stiffness of the structural material are not known accurately. For this reason it is adequate to analyse certain structures as if they are one- or two-dimensional. Thus the engineer's theory of beams is one-dimensional: the distribution of direct and shearing stresses across any section is assumed to depend only on the moment and shear at that section. By the same token, a plate, which is characterized by the fact that its thickness is small compared with its other linear dimensions, may be analysed in a two-dimensional manner. The simplest and most widely used plate theory is the *classical small-deflexion theory* which we will now consider.

The classical small-deflexion theory of plates, developed by Lagrange (1811), is based on the following assumptions:

(i) points which lie on a normal to the mid-plane of the undeflected plate lie on a normal to the mid-plane of the deflected plate;
(ii) the stresses normal to the mid-plane of the plate, arising from the applied loading, are negligible in comparison with the stresses in the plane of the plate;
(iii) the slope of the deflected plate in any direction is small so that its square may be neglected in comparison with unity;
(iv) the mid-plane of the plate is a 'neutral plane', that is, any mid-plane stresses arising from the deflexion of the plate into a non-developable surface may be ignored.

These assumptions have their counterparts in the engineer's theory of beams; assumption (i), for example, corresponds to the dual assumptions in beam theory that 'plane sections remain plane' and 'deflexions due to shear may be neglected'.

Possible sources of error arising from these assumptions are discussed later.

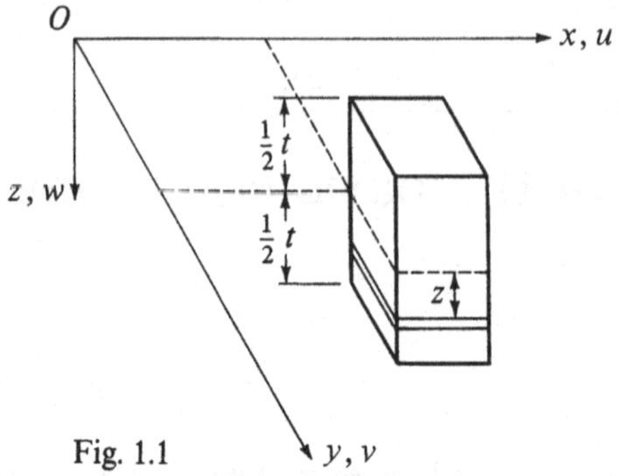

Fig. 1.1

1.1 Stress–strain relations

Let us consider now the state of stress in a plate with an arbitrary small deflexion $w(x, y)$ (see Fig. 1.1). The mid-plane is a neutral plane and accordingly we shall focus attention on the state of strain, and hence the state of stress, in a plane at a distance z from the mid-plane. The slopes of the mid-plane are $(\partial w/\partial x)$ and $(\partial w/\partial y)$ so that the displacements u and v in the x, y-plane at a distance z from the mid-plane are given by

$$\left.\begin{aligned} u &= -z\frac{\partial w}{\partial x} \\ v &= -z\frac{\partial w}{\partial y}. \end{aligned}\right\} \tag{1.1}$$

The strains in this x, y-plane are therefore given by

$$\left.\begin{aligned} \varepsilon_x &= \frac{\partial u}{\partial x} \\ &= -z\frac{\partial^2 w}{\partial x^2}, \\ \varepsilon_y &= \frac{\partial v}{\partial y} \\ &= -z\frac{\partial^2 w}{\partial y^2}, \\ \varepsilon_{xy} &= \frac{\partial u}{\partial y}+\frac{\partial v}{\partial x} \\ &= -2z\frac{\partial^2 w}{\partial x\,\partial y}. \end{aligned}\right\} \tag{1.2}$$

Now by virtue of assumption (ii) of, a state of plane stress exists in the x, y-plane so that the strains $\varepsilon_x, \varepsilon_y, \varepsilon_{xy}$ are related to the stresses $\sigma_x, \sigma_y, \tau_{xy}$ by the relations

$$\left.\begin{aligned}
\varepsilon_x &= \frac{1}{E}(\sigma_x - \nu\sigma_y)\\
\varepsilon_y &= \frac{1}{E}(\sigma_y - \nu\sigma_x)\\
\varepsilon_{xy} &= \frac{1}{G}\tau_{xy} = \frac{2(1+\nu)}{E}\tau_{xy}.
\end{aligned}\right\} \tag{1.3}$$

Equations (1.2) and (1.3) may be combined to give

$$\left.\begin{aligned}
\sigma_x &= -\frac{Ez}{1-\nu^2}\left(\frac{\partial^2 w}{\partial x^2} + \nu\frac{\partial^2 w}{\partial y^2}\right)\\
\sigma_y &= -\frac{Ez}{1-\nu^2}\left(\frac{\partial^2 w}{\partial y^2} + \nu\frac{\partial^2 w}{\partial x^2}\right)\\
\tau_{xy} &= -\frac{Ez}{1+\nu}\frac{\partial^2 w}{\partial x \partial y}.
\end{aligned}\right\} \tag{1.4}$$

These stresses vary linearly through the thickness of the plate and are equivalent to moments per unit length acting on an element of the plate, as shown in Fig. 1.2. Thus,

$$\left.\begin{aligned}
M_x &= \int_{-\frac{1}{2}t}^{\frac{1}{2}t} z\sigma_x \mathrm{d}z\\
&= -D\left(\frac{\partial^2 w}{\partial x^2} + \nu\frac{\partial^2 w}{\partial y^2}\right),\\
M_y &= \int_{-\frac{1}{2}t}^{\frac{1}{2}t} z\sigma_y \mathrm{d}z\\
&= -D\left(\frac{\partial^2 w}{\partial y^2} + \nu\frac{\partial^2 w}{\partial x^2}\right),\\
M_{xy} &= \int_{-\frac{1}{2}t}^{\frac{1}{2}t} z\tau_{xy} \mathrm{d}z\\
&= -D(1-\nu)\frac{\partial^2 w}{\partial x \partial y},
\end{aligned}\right\} \tag{1.5}$$

where the *flexural rigidity* D of the plate is defined by

$$D = \frac{Et^3}{12(1-\nu^2)}. \tag{1.6}$$

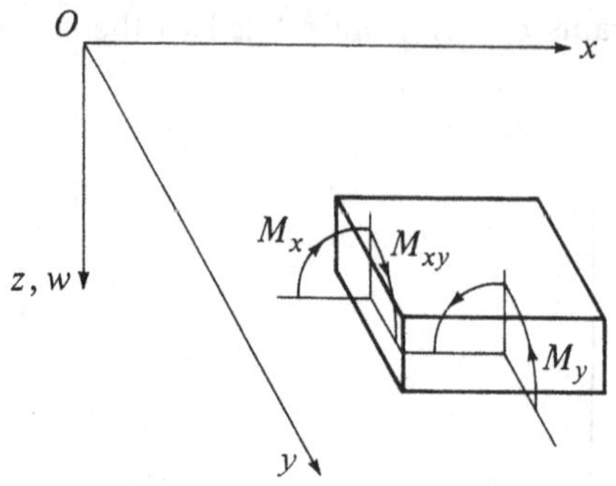

Fig. 1.2

Note, too, that because of the equality of complementary shears τ_{xy} and τ_{yx} it follows that

$$M_{xy} = M_{yx}. \tag{1.7}$$

In what follows, no distinction is drawn between the complementary twisting moments, and the symbol M_{xy} is used to denote them both.

1.1.1 Curvatures of the deflected plate

The *curvatures* of the plate in planes parallel to the x, z- and y, z-planes are $-(\partial^2 w/\partial x^2)$ and $-(\partial^2 w/\partial y^2)$, and these will sometimes by represented by the symbols κ_x and κ_y. Strictly speaking, the curvature κ_x, for example, is given by

$$\kappa_x = \frac{-\dfrac{\partial^2 w}{\partial x^2}}{\left\{1 + \left(\dfrac{\partial w}{\partial x}\right)^2\right\}^{3/2}}, \tag{1.8}$$

but in virtue of assumption (iii) the denominator may be taken equal to unity. The minus sign has been introduced so that an increase in M_x causes an increase in κ_x. The term $-(\partial^2 w/\partial x \partial y)$ is the *twisting curvature* and is represented by the symbol κ_{xy}.

It can be seen from (1.5) that the curvature at a point may be expressed simply in terms of the moments per unit length. Thus we find

$$\left.\begin{aligned} \kappa_x &= (M_x - \nu M_y)/\{(1-\nu^2)D\}, \\ \kappa_y &= (M_y - \nu M_x)/\{(1-\nu^2)D\}, \\ \kappa_{xy} &= (1+\nu)M_{xy}/\{(1-\nu)D\}. \end{aligned}\right\} \tag{1.9}$$

1.1.2 Deflexion of plate with constant curvatures

It can be readily verified by differentiation that the deflected form of a

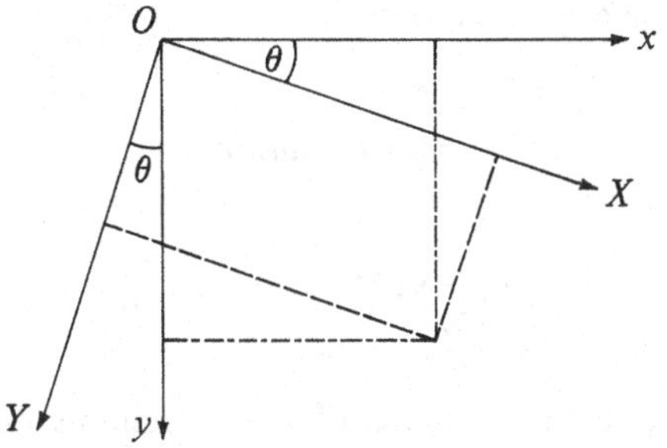

Fig. 1.3

plate under constant values of the curvatures κ_x, κ_y, κ_{xy} is given by

$$-w = \tfrac{1}{2}\kappa_x x^2 + \kappa_{xy}xy + \tfrac{1}{2}\kappa_y y^2, \tag{1.10}$$

to which may be added a rigid body displacement of the form $(Ax + By + C)$.

1.2 Rotation of axes of reference

If the deflected form of (1.10) is referred to a new system of Cartesian axes OX, OY obtained by rotating the axes Ox, Oy through an angle θ, as in Fig. 1.3, so that

$$\left.\begin{aligned} x &= X\cos\theta - Y\sin\theta \\ y &= X\sin\theta + Y\cos\theta, \end{aligned}\right\} \tag{1.11}$$

it is seen that

$$\begin{aligned} -w &= \tfrac{1}{2}\kappa_x(X\cos\theta - Y\sin\theta)^2 + \kappa_{xy}(X\cos\theta - Y\sin\theta) \\ &\quad\cdot(X\sin\theta + Y\cos\theta) + \tfrac{1}{2}\kappa_y(X\sin\theta + Y\cos\theta)^2 \\ &= \tfrac{1}{2}(\kappa_x\cos^2\theta + 2\kappa_{xy}\sin\theta\cos\theta + \kappa_y\sin^2\theta)X^2 \\ &\quad + \{(\kappa_y - \kappa_x)\sin\theta\cos\theta + \kappa_{xy}(\cos^2\theta - \sin^2\theta)\}XY \\ &\quad + \tfrac{1}{2}(\kappa_x\sin^2\theta - 2\kappa_{xy}\sin\theta\cos\theta + \kappa_y\cos^2\theta)Y^2 \\ &\equiv \tfrac{1}{2}\kappa_X X^2 + \kappa_{XY}XY + \tfrac{1}{2}\kappa_Y Y^2. \end{aligned} \tag{1.12}$$

Equating coefficients of X^2, Y^2 and XY makes it possible to express $\kappa_X, \kappa_Y, \kappa_{XY}$ in terms of $\kappa_x, \kappa_y, \kappa_{xy}$:

$$\left.\begin{aligned} \kappa_X &= \kappa_x\cos^2\theta + 2\kappa_{xy}\sin\theta\cos\theta + \kappa_y\sin^2\theta \\ \kappa_Y &= \kappa_x\sin^2\theta - 2\kappa_{xy}\sin\theta\cos\theta + \kappa_y\cos^2\theta \\ \kappa_{XY} &= (\kappa_y - \kappa_x)\sin\theta\cos\theta + \kappa_{xy}(\cos^2\theta - \sin^2\theta). \end{aligned}\right\} \tag{1.13}$$

1.2.1 Invariant relationships

Two invariant relationships between the curvatures may be obtained from (1.13) by eliminating θ. For example, by adding the first two of (1.13) and dividing by 2 we find that for any value of θ

$$\tfrac{1}{2}(\kappa_X + \kappa_Y) = \tfrac{1}{2}(\kappa_x + \kappa_y). \tag{1.14}$$

This expression is referred to as the *average curvature.*

Similarly, it can be shown that

$$\kappa_{XY}^2 + \left(\frac{\kappa_Y - \kappa_X}{2}\right)^2 = \kappa_{xy}^2 + \left(\frac{\kappa_y - \kappa_x}{2}\right)^2. \tag{1.15}$$

This sum is later shown to be the square of the maximum twisting curvature. Furthermore, (1.14) and (1.15) may be combined to give

$$\kappa_X \kappa_Y - \kappa_{XY}^2 = \kappa_x \kappa_y - \kappa_{xy}^2. \tag{1.16}$$

This expression is referred to as the *Gaussian curvature.*

1.2.2 Principal axes of reference

It is frequently convenient to choose the angle θ in such a way that the twisting curvature κ_{XY} vanishes. Now the twisting curvature, given by (1.13), may be written in the form

$$\begin{aligned}\kappa_{XY} &= \tfrac{1}{2}(\kappa_y - \kappa_x)\sin 2\theta + \kappa_{xy}\cos 2\theta \\ &= \left\{\kappa_{xy}^2 + \left(\frac{\kappa_y - \kappa_x}{2}\right)^2\right\}^{1/2} \sin 2(\theta - \beta)\end{aligned} \tag{1.17}$$

where

$$\beta = \tfrac{1}{2}\tan^{-1}\left(\frac{2\kappa_{xy}}{\kappa_x - \kappa_y}\right), \tag{1.18}$$

and κ_{XY} therefore vanishes when

$$\theta = \beta \quad \text{or} \quad \tfrac{1}{2}\pi + \beta. \tag{1.19}$$

When the axes are chosen to satisfy (1.19) they are called *principal axes.* Now, corresponding to (1.17), we may write

$$\left.\begin{aligned}\kappa_X &= \tfrac{1}{2}(\kappa_x + \kappa_y) + \left\{\kappa_{xy}^2 + \left(\frac{\kappa_y - \kappa_x}{2}\right)^2\right\}^{1/2} \cos 2(\theta - \beta) \\ \kappa_Y &= \tfrac{1}{2}(\kappa_x + \kappa_y) - \left\{\kappa_{xy}^2 + \left(\frac{\kappa_y - \kappa_x}{2}\right)^2\right\}^{1/2} \cos 2(\theta - \beta)\end{aligned}\right\} \tag{1.20}$$

which shows that κ_X and κ_Y assume maximum and minimum values when (1.19) is satisfied.

Maximum twisting curvature. The maximum value of the twisting curvature, given by (1.17), occurs when

$$\sin 2(\theta - \beta) = \pm 1,$$

that is,

$$\theta = \pi/4 + \beta \quad \text{or} \quad 3\pi/4 + \beta \tag{1.21}$$

which shows that the twisting curvature is a maximum on planes bisecting the principal planes of curvature, and on such planes

$$\kappa_X = \kappa_Y = \tfrac{1}{2}(\kappa_x + \kappa_y). \tag{1.22}$$

1.2.3 Resolution of moments M_X, M_Y, M_{XY}

The moments M_X, M_Y, M_{XY} may be expressed in terms of M_x, M_y, M_{xy} either directly by considering the equilibrium of an element of the plate, or indirectly by virtue of (1.5) and (1.13). Thus we find

$$\left.\begin{aligned} M_X &= M_x \cos^2\theta + 2M_{xy}\sin\theta\cos\theta + M_y\sin^2\theta \\ M_Y &= M_x \sin^2\theta - 2M_{xy}\sin\theta\cos\theta + M_y\cos^2\theta \\ M_{XY} &= (M_y - M_x)\sin\theta\cos\theta + M_{xy}(\cos^2\theta - \sin^2\theta). \end{aligned}\right\} \tag{1.23}$$

These equations have the same form as those of (1.13) for the curvatures. Furthermore, from (1.5)

$$\frac{\kappa_{xy}}{\kappa_x - \kappa_y} = \frac{M_{xy}}{M_x - M_y}$$

and equations (1.14)–(1.22) are therefore valid when the symbol κ is replaced throughout by M. Thus we find,

$$\left.\begin{aligned} & M_x + M_y = M_X + M_Y, \\ & \textit{maximum twisting moment} = \left\{ M_{xy}^2 + \left(\frac{M_y - M_x}{2}\right)^2 \right\}^{1/2} \\ & \textit{principal moments} = \tfrac{1}{2}(M_x + M_y) \pm \left\{ M_{xy}^2 + \left(\frac{M_y - M_x}{2}\right)^2 \right\}^{1/2}. \end{aligned}\right\} \tag{1.24}$$

1.3 Equilibrium

A typical element of the plate bounded by the lines x, $x + \delta x$, y, $y + \delta y$ may be subjected to a distributed normal loading of intensity q, positive if acting in the direction of positive w. The resultant normal force $q\delta x\delta y$ on the plate elements is reacted by normal shears acting over the sides of the element. The magnitude per unit length of the shears acting on a side normal to the x-axis is denoted by Q_x, that on a side normal to the y-axis by Q_y (see Fig. 1.4). Resolving normal to the plate gives

$$\frac{\partial Q_x}{\partial x}\delta x\delta y + \frac{\partial Q_y}{\partial y}\delta x\delta y + q\delta x\delta y = 0,$$

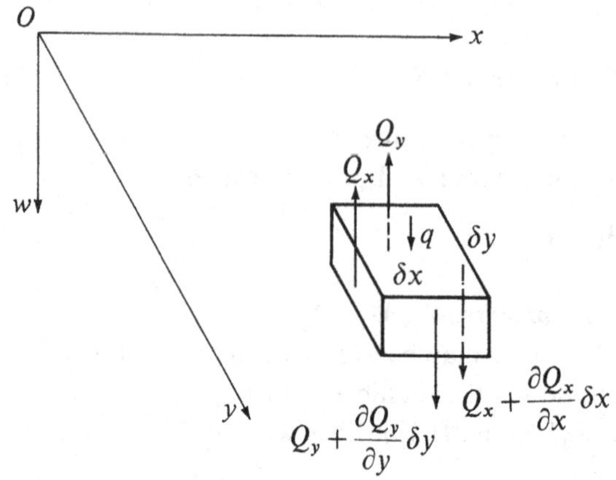

Fig. 1.4

whence

$$\frac{\partial Q_x}{\partial x} + \frac{\partial Q_y}{\partial y} + q = 0. \tag{1.25}$$

Similarly, by taking moments about the y- and x-axes of all forces acting on the element, we obtain

$$\left.\begin{aligned} Q_x &= \frac{\partial M_x}{\partial x} + \frac{\partial M_{xy}}{\partial y} \\ Q_y &= \frac{\partial M_y}{\partial y} + \frac{\partial M_{xy}}{\partial x}. \end{aligned}\right\} \tag{1.26}$$

An equation of equilibrium may now be expressed in terms of derivatives of the moments and the applied loading by eliminating Q_x, Q_y from (1.25) and (1.26), giving

$$\frac{\partial^2 M_x}{\partial x^2} + 2\frac{\partial^2 M_{xy}}{\partial x \partial y} + \frac{\partial^2 M_y}{\partial y^2} + q = 0. \tag{1.27}$$

1.4 Differential equation for the deflexion

The differential equation for the deflexion of the plate is obtained by substituting the moment–curvature relationships of (1.5) in the equilibrium equation (1.27). If the flexural rigidity D is a function of x, y we obtain

$$\frac{\partial^2}{\partial x^2}\left\{D\left(\frac{\partial^2 w}{\partial x^2} + \nu\frac{\partial^2 w}{\partial y^2}\right)\right\} + 2(1-\nu)\frac{\partial^2}{\partial x\,\partial y}\left\{D\frac{\partial^2 w}{\partial x\,\partial y}\right\}$$
$$+\frac{\partial^2}{\partial y^2}\left\{D\left(\frac{\partial^2 w}{\partial y^2} + \nu\frac{\partial^2 w}{\partial x^2}\right)\right\} = q \tag{1.28a}$$

which may be written in the following invariant form,

$$\nabla^2(D\nabla^2 w) - (1-\nu)\Diamond^4(D, w) = q, \tag{1.28b}$$

where

$$\nabla^2 = \frac{\partial^2}{\partial x^2} + \frac{\partial^2}{\partial y^2},$$

and the "die-operator" is defined by

$$\begin{aligned}\Diamond^2(D, w) &\equiv \tfrac{1}{2}\{(\nabla^2 D)(\nabla^2 w) + \nabla^2(D\nabla^2 w + w\nabla^2 D)\}\\ &\quad - \tfrac{1}{4}\{\nabla^2(Dw) + D\nabla^4 w + w\nabla^4 D\}\\ &\equiv \frac{\partial^2 D}{\partial x^2}\frac{\partial^2 w}{\partial y^2} - 2\frac{\partial^2 D}{\partial x \partial y}\frac{\partial^2 w}{\partial x \partial y} + \frac{\partial^2 D}{\partial y^2}\frac{\partial^2 w}{\partial x^2}.\end{aligned}$$

1.4.1 Plate with constant rigidity

When D is a constant, (1.28) simplifies to

$$\nabla^4 w = \frac{q}{D} \tag{1.29}$$

and the shears per unit length, Q_x and Q_y, may be expressed in the form

$$\left.\begin{aligned} Q_x &= -D\frac{\partial}{\partial x}\nabla^2 w\\ Q_y &= -D\frac{\partial}{\partial y}\nabla^2 w.\end{aligned}\right\} \tag{1.30}$$

Reduction to two harmonic equations

It was shown in Section 1.2.3 that $(M_x + M_y)$ was invariant with respect to the orientation of axes, and equal to the sum of the principal moments. Another representation of this invariance is obtained from (1.5), which yields

$$M_x + M_y = -(1+\nu)D\nabla^2 w.$$

Thus, if we write, say,

$$(M_x + M_y)/(1+\nu) = M$$

equation (1.29) can be expressed as two simultaneous harmonic equations, namely,

$$\left.\begin{aligned}\nabla^2 M &= -q,\\ \nabla^2 w &= -M/D.\end{aligned}\right\} \tag{1.31}$$

This representation was introduced by Marcus (1932). It is particularly

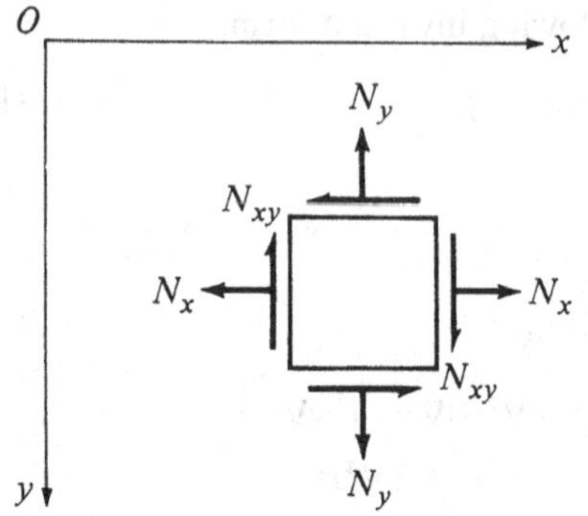

Fig. 1.5

useful in the case of simply supported plates of polygonal shape because the term M then vanishes along the boundary and thus allows the two equations in (1.31) to be integrated in succession. By employing a variational technique discussed in Chapter 6, Morley (1963) has used these equations to find solutions for plates whose boundaries are clamped.

1.5 Effect of forces in the plane of the plate

The differential equation (1.28) governing the deflexion of the plate is based on the tacit assumption that the mid-plane of the plate is free from stress, so that there is no resultant force in the plane of the plate. Resultant forces in the plane of the plate are referred to as *middle-surface forces.* They can arise directly owing to the application of middle-surface forces at the boundary or indirectly due to variations in temperature as discussed in Section 1.6. Middle-surface forces may also arise owing to straining of the mid-plane of the plate when it deflects into a non-developable surface, but this is a large-deflexion effect which is considered in Part II; in the small-deflexion régime considered here, such straining is of secondary importance as it varies, roughly speaking, as the square of the deflexion.

Consider now the effect of middle-surface forces per unit length, N_x, N_y, N_{xy}, as shown in Fig. 1.5. The distribution of these forces throughout the plate depends upon (i) their values along the boundary, (ii) the preservation of equilibrium in the plane of the plate, and (iii) compatibility of strains in the mid-plane of the plate. In many practical cases the plate thickness is constant and the forces are distributed along the boundary in such a way that N_x, N_y, N_{xy} maintain values that can be written down by inspection. But this is not always so, and we summarize below the equations needed to determine the distribution of these forces in the general case.

Equilibrium of an element $\delta x \delta y$ in the plane of the plate yields the conditions

$$\left.\begin{aligned}\frac{\partial N_x}{\partial x}+\frac{\partial N_{xy}}{\partial y}&=0\\ \frac{\partial N_y}{\partial y}+\frac{\partial N_{xy}}{\partial x}&=0.\end{aligned}\right\} \tag{1.32}$$

By the introduction of a *force function* these two conditions may be reduced to one, namely, that the forces per unit length may be derived from a single function Φ by double differentiation:

$$\left.\begin{aligned}N_x&=\frac{\partial^2\Phi}{\partial y^2},\\ N_y&=\frac{\partial^2\Phi}{\partial x^2},\\ N_{xy}&=-\frac{\partial^2\Phi}{\partial x\partial y}.\end{aligned}\right\} \tag{1.33}$$

If straining of the mid-plane of the plate due to deflexion of the plate is ignored, the strains in the mid-plane are related to the forces per unit length as follows:

$$\left.\begin{aligned}\frac{\partial u}{\partial x}&=(N_x-\nu N_y)/Et\\ \frac{\partial v}{\partial y}&=(N_y-\nu N_x)/Et\\ \frac{\partial u}{\partial y}+\frac{\partial v}{\partial x}&=N_{xy}/Gt.\end{aligned}\right\} \tag{1.34}$$

Now the left-hand sides of (1.34) satisfy the following differential identity

$$\frac{\partial^2}{\partial y^2}\left(\frac{\partial u}{\partial x}\right)+\frac{\partial^2}{\partial x^2}\left(\frac{\partial v}{\partial y}\right)-\frac{\partial^2}{\partial x\partial y}\left(\frac{\partial u}{\partial y}+\frac{\partial v}{\partial x}\right)\equiv 0 \tag{1.35}$$

which is referred to as the *condition of compatibility*. Expressed in terms of the right-hand sides of (1.34), this condition becomes

$$\frac{\partial^2}{\partial y^2}\left(\frac{N_x-\nu N_y}{t}\right)+\frac{\partial^2}{\partial x^2}\left(\frac{N_y-\nu N_x}{t}\right)-2(1+\nu)\frac{\partial^2}{\partial x\partial y}\left(\frac{N_{xy}}{t}\right)=0 \tag{1.36}$$

which may be combined with (1.33) and rearranged in the following invariant form:

$$\nabla^2\left(\frac{1}{t}\nabla^2\Phi\right)-(1+\nu)\Diamond^4\left(\frac{1}{t},\Phi\right)=0 \tag{1.37}$$

where the operator $\Diamond^4(\ ,\ )$ is the same as that used in the derivation of (1.29). When t is constant, (1.37) reduces to

$$\nabla^4\Phi=0. \tag{1.38}$$

Let us assume now that the forces per unit length, N_x, N_y, N_{xy}, are known (or that the function Φ is known) and consider the equilibrium of an element $\delta x\delta y$ normal to the original plane of the plate. The component of force in the z-direction acting on the face whose coordinate is x is

$$-N_x\frac{\partial w}{\partial x}\delta y-N_{xy}\frac{\partial w}{\partial y}\delta y$$

and the component acting on the face at $x+\delta x$ is therefore

$$\left(N_x\frac{\partial w}{\partial x}+N_{xy}\frac{\partial w}{\partial y}\right)\delta y+\frac{\partial}{\partial x}\left(N_x\frac{\partial w}{\partial x}+N_{xy}\frac{\partial w}{\partial y}\right)\delta y\delta x.$$

There are similar components, with the symbols x and y interchanged, acting on the other faces. The resultant force in the z-direction acting on the element $\delta x\delta y$ is thus given by

$$\frac{\partial}{\partial x}\left(N_x\frac{\partial w}{\partial x}+N_{xy}\frac{\partial w}{\partial y}\right)\delta x\delta y+\frac{\partial}{\partial y}\left(N_y\frac{\partial w}{\partial y}+N_{xy}\frac{\partial w}{\partial x}\right)\delta x\delta y$$

$$=\left(N_x\frac{\partial^2 w}{\partial x^2}+2N_{xy}\frac{\partial^2 w}{\partial x\partial y}+N_y\frac{\partial^2 w}{\partial y^2}\right)\delta x\delta y. \tag{1.39}$$

A comparison with the result of Section 1.3 shows that the effect of middle-surface forces on the deflexion is equivalent to an additional pressure q',

$$q'=N_x\frac{\partial^2 w}{\partial x^2}+2N_{xy}\frac{\partial^2 w}{\partial x\partial y}+N_y\frac{\partial^2 w}{\partial y^2} \tag{1.40}$$

$$=\frac{\partial^2\Phi}{\partial y^2}\frac{\partial^2 w}{\partial x^2}-2\frac{\partial^2\Phi}{\partial x\partial y}\frac{\partial^2 w}{\partial x\partial y}+\frac{\partial^2\Phi}{\partial x^2}\frac{\partial^2 w}{\partial y^2}$$

$$=\Diamond^4(\Phi,w), \tag{1.40a}$$

by virtue of (1.33).

Equations (1.29) and (1.40a) may be combined to give the differential equation for the deflexion of a plate of varying thickness, including effects of middle-surface forces:

$$\nabla^2(D\nabla^2 w) - (1-\nu)\lozenge^4(D, w) = q + \lozenge^4(\Phi, w) \tag{1.41}$$

where the force function Φ satisfies (1.37).

Fortunately, we are generally concerned with plates of constant thickness subject to a known distribution of middle-surface forces, for which (1.30) and (1.40) yield

$$D\nabla^4 w = q + N_x \frac{\partial^2 w}{\partial x^2} + 2N_{xy}\frac{\partial^2 w}{\partial x \partial y} + N_y \frac{\partial^2 w}{\partial y^2}. \tag{1.42}$$

1.5.1 Plate on an elastic foundation

If a plate rests on an elastic foundation such that the restoring pressure is everywhere proportional to the deflexion, the resultant pressure acting on the plate assumes the form

$$q_{\text{res}} = q - kw, \tag{1.43}$$

where k is the *foundation modulus*.

The differential equation for the deflexion of the plate is obtained from the preceding analysis by substituting q_{res} for q. In particular, for the plate of constant thickness under the action of middle-surface forces, we obtain

$$D\nabla^4 w = q - kw + N_x \frac{\partial^2 w}{\partial x^2} + 2N_{xy}\frac{\partial^2 w}{\partial x \partial y} + N_y \frac{\partial^2 w}{\partial y^2}. \tag{1.44}$$

1.5.2 Vibration of a plate

When a plate is loaded statically, the *elastic reaction* of the plate is everywhere equal and opposite to the applied loading q. If there is no external applied loading but the plate is vibrating, the elastic reaction acting on each element of the plate (measured in the direction of negative w) produces an acceleration of each element of the plate in the same direction. The magnitude of the elastic reaction is thus equal to $-m(x, y)\partial^2 w/\partial t^2$, where $m(x, y)$ is the mass per unit area of the plate. The differential equation for the deflexion of the plate may now be obtained from the preceding analysis by substituting $-m\partial^2 w/\partial t^2$ for q. In particular, for a plate of constant thickness on an elastic foundation and under the action of middle-surface forces, we obtain

$$D\nabla^4 w + kw + m\frac{\partial^2 w}{\partial t^2} = N_x \frac{\partial^2 w}{\partial x^2} + 2N_{xy}\frac{\partial^2 w}{\partial x \partial y} + N_y \frac{\partial^2 w}{\partial y^2} \tag{1.45}$$

in which $w = w(x, y, t)$.

In many problems associated with vibrations of plates we are concerned with the vibration in one particular *mode* characterized by each element of the plate executing simple harmonic motion in phase with all other elements. Thus we may write

$$w(x, y, t) = w(x, y) \sin\{\Omega(t - t_0)\}, \tag{1.46}$$

where Ω is the circular frequency.

Substitution of (1.46) in (1.45) and division throughout by $\sin\{\Omega(t-t_0)\}$ then yields the equation

$$D\nabla^4 w + (k - m\Omega^2)w = N_x \frac{\partial^2 w}{\partial x^2} + 2N_{xy}\frac{\partial^2 w}{\partial x \partial y} + N_y \frac{\partial^2 w}{\partial y^2}. \tag{1.47}$$

The fact that k and $m\Omega^2$ occur only in the combination $(k - m\Omega^2)$ implies that any mode for a plate for which k is zero, say $w_1(x, y)$ and Ω_1, will also be appropriate to a similar plate on an elastic foundation, but the frequency Ω_2, say, is increased according to the relation

$$\Omega_2 = (\Omega_1^2 + k/m)^{1/2}. \tag{1.48}$$

1.6 Thermal stress effects

In general, when a plate is heated by conduction, radiation or convection the temperatures in the plate vary slowly in comparison with the natural periods of vibration of the plate. For this reason the plate may be analysed in a quasi-static manner. Further, unless there are abrupt changes in the distribution of the surface temperature – as, for example, near a spot-weld – the basic assumption of plate theory may be retained, namely, that points which lie on a normal to the mid-plane of the undeflected plate lie on a normal to the mid-plane of the deflected plate. Consider therefore a plate whose temperature $T(x, y, z)$, measured from some convenient datum such as room temperature, varies in an arbitrary but sufficiently smooth manner in the plane of the plate and through the thickness. We first focus attention on a typical element $\delta x \delta y$ and we determine the forces and moments per unit length required to prevent any planar displacement or rotation at the boundaries of the element. In the x, y-plane passing through the point z in the plate element, the planar stresses satisfy the equations

$$\left.\begin{aligned} \partial u/\partial x = \alpha T(x, y, z) + (\sigma_x - \nu\sigma_y)/E = 0, \\ \text{and similarly} \qquad\qquad \\ \alpha T(x, y, z) + (\sigma_y - \nu\sigma_x)/E = 0, \\ \tau_{xy}/G = 0, \end{aligned}\right\} \tag{1.49}$$

where α is the coefficient of thermal expansion which may vary with the temperature. It follows that

$$\sigma_x = \sigma_y = -\frac{E}{1-\nu}\alpha T(x, y, z). \tag{1.50}$$

The above relations are also valid if ν and E vary with temperature, but in what follows we confine attention to materials in which ν can be assumed to be constant; a temperature-dependent Young modulus is denoted by E_T.

At this point we note that with a varying value of the Young modulus the neutral surface does not necessarily pass through the mid-thickness ($z = 0$) but passes through $z = z^*$, where

$$z^* = \int_{-\frac{1}{2}t}^{\frac{1}{2}t} zE_T \, dz \bigg/ \int_{-\frac{1}{2}t}^{\frac{1}{2}t} E_T \, dz. \tag{1.51}$$

Likewise, the plate stiffness per unit length in the plane of the neutral surface, Et, if E is constant, is given by

$$S_T, \quad \text{say,} \quad = \int_{-\frac{1}{2}t}^{\frac{1}{2}t} E_T \, dz, \tag{1.52}$$

and the flexural rigidity is given by

$$D_T, \quad \text{say,} \quad = \frac{1}{(1-\nu^2)} \int_{-\frac{1}{2}t}^{\frac{1}{2}t} (z - z^*)^2 E_T \, dz. \tag{1.53}$$

Referring back to (1.50), we can now integrate the stresses σ_x, σ_y through the thickness of the plate to yield the following resultant forces in the plane of the neutral surface and moments per unit length about the neutral surface:

$$N_x = N_y = -\frac{1}{1-\nu} \int_{-\frac{1}{2}t}^{\frac{1}{2}t} E_T \alpha T(x, y, z) \, dz, \tag{1.54}$$

and

$$M_x = M_y = -\frac{1}{1-\nu} \int_{-\frac{1}{2}t}^{\frac{1}{2}t} (z - z^*) E_T \alpha T(x, y, z) \, dz. \tag{1.55}$$

The above equations refer to an element whose boundaries are restrained against any rotation or planar displacement. If the boundaries are free from such restraint, so that there are no force or moment resultants per unit length, equal and opposite values of N_x, M_x and so on must be superimposed on the above stress system. It follows that for an *unconstrained* element satisfying the basic assumption of plate theory the direct strains *in the neutral surface* are given by

$$\varepsilon_x^* = \varepsilon_y^* = \varepsilon_T, \quad \text{say,}$$

$$= \int_{-\frac{1}{2}t}^{\frac{1}{2}t} E_T \alpha T(x, y, z) dz \bigg/ \int_{-\frac{1}{2}t}^{\frac{1}{2}t} E_T \, dz, \tag{1.56}$$

and the curvatures are given by

$$\kappa_x^* = \kappa_y^* = \kappa_T, \quad \text{say,}$$
$$= \int_{-\frac{1}{2}t}^{\frac{1}{2}t} (z - z^*) E_T \alpha T(x, y, z)\, dz \Big/ \int_{-\frac{1}{2}t}^{\frac{1}{2}t} (z - z^*)^2 E_T\, dz. \tag{1.57}$$

The thermal strain ε_T and curvature κ_T thus specify the overall effect of any variation of temperature through the thickness.

When the variation of $T(x, y, z)$ and E_T is such that the neutral surface $(z = z^*)$ varies over the plate, the effect is similar to a plate with an initial deflexion $w_o(x, y)$, where

$$w_o(x, y) = z^*(x, y). \tag{1.58}$$

The behaviour of such plates in the presence of resultant planar forces requires large-deflexion theory (see Part II). However, when the range of temperature is such that E_T can be assumed to be constant, that is, $E_T = E$, the neutral surface is at the mid-thickness, and small-deflexion theory is valid.

1.6.1 Young modulus independent of temperature

For such plates, $z^* = 0$ and (1.56), (1.57) simplify to

$$\left.\begin{aligned} \varepsilon_T &= \frac{1}{t} \int_{-\frac{1}{2}t}^{\frac{1}{2}t} \alpha T(x, y, z)\, dz, \\ \kappa_T &= \frac{12}{t^3} \int_{-\frac{1}{2}t}^{\frac{1}{2}t} z \alpha T(x, y, z)\, dz. \end{aligned}\right\} \tag{1.59}$$

General strain and curvature relations

When there are middle-surface forces the middle-surface strains are given by

$$\left.\begin{aligned} \frac{\partial u}{\partial x} &= \varepsilon_T + (N_x - \nu N_y)/Et, \\ \frac{\partial v}{\partial y} &= \varepsilon_T + (N_y - \nu N_x)/Et, \\ \frac{\partial u}{\partial y} + \frac{\partial v}{\partial x} &= N_{xy}/Gt. \end{aligned}\right\} \tag{1.60}$$

Referring now to (1.35), we see that the condition of compatibility can be expressed in the following invariant form:

$$\nabla^2\left(\frac{1}{t}\nabla^2\Phi\right) - (1 + \nu)\Diamond^4\left(\frac{1}{t}, \Phi\right) + E\nabla^2\varepsilon_T = 0. \tag{1.61}$$

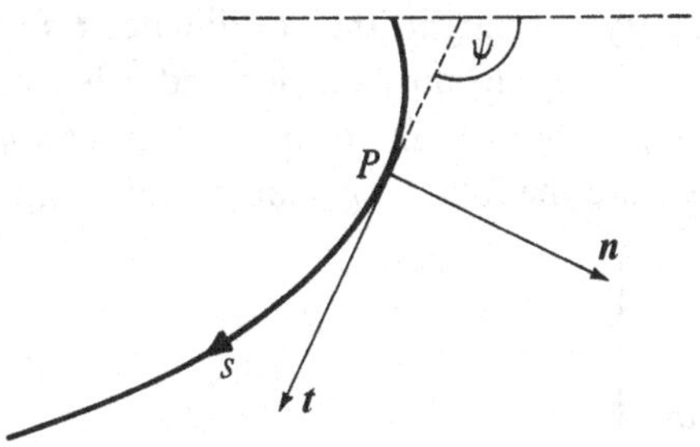

Fig. 1.6

Similarly, the moment–curvature relations, see (1.9), are now

$$\left.\begin{aligned} \kappa_x &= \kappa_T + (M_x - \nu M_y)/\{(1-\nu^2)D\}, \\ \kappa_y &= \kappa_T + (M_y - \nu M_x)/\{(1-\nu^2)D\}, \\ \kappa_{xy} &= M_{xy}/\{(1-\nu)D\}, \end{aligned}\right\} \tag{1.62}$$

and it follows that the equilibrium equation, including effects of middle-surface forces, can be expressed in the form

$$\nabla^2(D\nabla^2 w) - (1-\nu)\lozenge^4(D, w) + (1+\nu)\nabla^2(D\kappa_T) = q + \lozenge^4(\Phi, w). \tag{1.63}$$

For plates of constant thickness, (1.61) and (1.63) reduce to

$$\nabla^4\Phi + Et\nabla^2\varepsilon_T = 0, \tag{1.64}$$

and

$$D\nabla^4 w + (1+\nu)D\nabla^2\kappa_T = q + \lozenge^4(\Phi, w). \tag{1.65}$$

Note, finally, that for plates of constant thickness small-deflexion theory is also valid when the Young modulus varies with temperature but the temperature does not vary in the plane of the plate, that is, $T = T(z)$. For such plates, ε_T and κ_T are constants for a given variation of $T(z)$ but, more important, so too are z^*, S_T and D_T.

1.7 General boundary conditions

The partial differential equation governing the deflexion of a plate is of the fourth order. It follows that along the boundary of the plate two conditions (and only two) are required if w is to be uniquely determined.

Typical boundary conditions for a plate of arbitrary shape and variable rigidity are expressed here in terms of the deflexion w and its derivatives. Boundary conditions involving the twisting moments per unit length require special attention. Temperature effects are discussed in Section 1.7.6.

Let $\boldsymbol{n}, \boldsymbol{t}$ be measured along the outward normal and tangent to the edge at a typical point P, as shown in Fig. 1.6, and let s be measured along the boundary. If the boundary is straight, the coordinates $\boldsymbol{t}, s$ coincide, but if the boundary is curved, they do not coincide and it is convenient to express certain derivatives of w in terms of $\boldsymbol{n}, s$ rather than $\boldsymbol{n}, \boldsymbol{t}$. The relations which are required are the following geometrical identities

$$\left.\begin{aligned} \frac{\partial w}{\partial \boldsymbol{t}} &\equiv \frac{\partial w}{\partial s} \\ \frac{\partial^2 w}{\partial \boldsymbol{t}^2} &\equiv \frac{\partial^2 w}{\partial s^2} + \frac{\partial \psi}{\partial s}\frac{\partial w}{\partial \boldsymbol{n}} \\ \frac{\partial^2 w}{\partial \boldsymbol{n}\partial \boldsymbol{t}} &\equiv \frac{\partial^2 w}{\partial \boldsymbol{n}\partial s} - \frac{\partial \psi}{\partial s}\frac{\partial w}{\partial s} \end{aligned}\right\} \tag{1.66}$$

in which $\partial\psi/\partial s$ is the curvature of the boundary.

1.7.1 *Clamped edge*

Along the boundary, the deflexion and slope normal to the boundary are zero, so that

$$w = 0, \tag{1.67}$$

and

$$\frac{\partial w}{\partial \boldsymbol{n}} = 0. \tag{1.68}$$

1.7.2 *Simply supported edge*

Along the boundary, the deflexion and the moment per unit length, M_n, are zero, so that

$$w = 0 \tag{1.69}$$

and, from (1.5),

$$\frac{\partial^2 w}{\partial \boldsymbol{n}^2} + \nu\frac{\partial^2 w}{\partial \boldsymbol{t}^2} = 0$$

which may be rewritten, using (1.66) and (1.69), in the form

$$\frac{\partial^2 w}{\partial \boldsymbol{n}^2} + \nu\frac{\partial \varphi}{\partial s}\frac{\partial w}{\partial \boldsymbol{n}} = 0. \tag{1.70}$$

If the boundary is straight, (1.70) reduces to

$$\frac{\partial^2 w}{\partial \boldsymbol{n}^2} = 0. \tag{1.71}$$

1.7.3 *Edge elastically supported against rotation*

One of the conditions at an edge may be such that a rotation $\partial w/\partial \boldsymbol{n}$ of the plate is resisted by a moment $\chi(\partial w/\partial \boldsymbol{n})$, say, owing to the surrounding

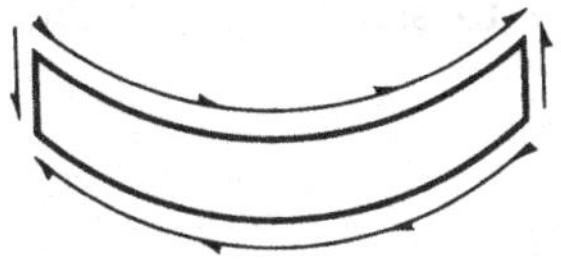

Fig. 1.7

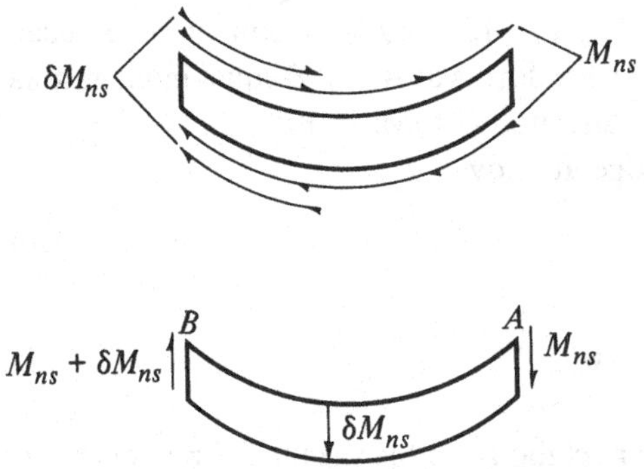

Fig. 1.8

structure. Such a condition is intermediate between clamped and simply supported, and is given by

$$\left(\frac{\chi}{D}+\nu\frac{\partial\psi}{\partial s}\right)\frac{\partial w}{\partial n}+\frac{\partial^2 w}{\partial n^2}=0. \tag{1.72}$$

1.7.4 Free edge

For a real plate, we require the vanishing at the boundary of M_n, M_{ns} and Q_n, that is, one condition more than is mathematically feasible for the classical theory of plates. To overcome this apparent difficulty, we must refer again to assumption (i) which states, in effect, that the plate is rigid in shear. Consider now the action of a vanishingly narrow strip along the boundary. Such a strip is rigid in shear, though perfectly flexible in bending, and resists without deformation the shear loading shown in Fig. 1.7.

The horizontal components of the shear loading in Fig. 1.7 are statically equivalent to a constant value of M_{ns} along the boundary; the action of this vanishingly narrow strip can therefore be regarded as converting such a 'horizontal' loading into equal and opposite vertical forces, equal in magnitude to M_{ns}, acting at the ends of the strip, as shown in Fig. 1.8. There is no need to limit the argument to a constant value of M_{ns}; if the boundary twisting moment increases by an amount $(\partial M_{ns}/\partial s)\delta s$ over a distance δs, the action of the vanishingly narrow strip is to convert this into vertical forces which are equivalent to shears Q_n' per unit length where

$$Q_n' = \frac{\partial M_{ns}}{\partial s}. \tag{1.73}$$

Thus we have shown that *within the framework of the classical theory of plates* no distinction can be drawn between an edge twisting moment M_{ns} which varies, say, from M_{ns}^{A} to M_{ns}^{B}, and a system of edge shears given by (1.73) together with vertical forces at A and B equal in magnitude to M_{ns}^{A} and M_{ns}^{B}, acting in the directions shown in Fig. 1.8.

The conditions for a free edge are now

$$M_n = 0 \tag{1.74}$$

and

$$Q_n + Q_n' = Q_n + \frac{\partial M_{ns}}{\partial s} = 0 \tag{1.75}$$

which is now a joint requirement embodying the shears and rate of change of twisting moment.

This joint requirement was first derived by Kirchhoff (1850) from variational considerations, and the underlying physical explanation was given by Kelvin and Tait (1883). Expressed in terms of the deflexion, (1.74) becomes

$$D\left\{\frac{\partial^2 w}{\partial n^2} + \nu\left(\frac{\partial^2 w}{\partial s^2} + \frac{\partial \psi}{\partial s}\frac{\partial w}{\partial n}\right)\right\} = 0 \tag{1.76}$$

while (1.75) becomes

$$\begin{aligned} &D\left\{\frac{\partial}{\partial n}\nabla^2 w + (1-\nu)\frac{\partial}{\partial s}\left(\frac{\partial^2 w}{\partial n\,\partial s} - \frac{\partial \psi}{\partial s}\frac{\partial w}{\partial s}\right)\right\} \\ &\quad + \frac{\partial D}{\partial n}\left\{\frac{\partial^2 w}{\partial n^2} + \nu\left(\frac{\partial^2 w}{\partial s^2} + \frac{\partial \psi}{\partial s}\frac{\partial w}{\partial n}\right)\right\} \\ &\quad + 2(1-\nu)\frac{\partial D}{\partial s}\left(\frac{\partial^2 w}{\partial n\,\partial s} - \frac{\partial \psi}{\partial s}\frac{\partial w}{\partial s}\right) = 0 \end{aligned} \tag{1.77}$$

which reduces to

$$\frac{\partial}{\partial n}\nabla^2 w + (1-\nu)\frac{\partial}{\partial s}\left(\frac{\partial^2 w}{\partial n\,\partial s} - \frac{\partial \psi}{\partial s}\frac{\partial w}{\partial s}\right) = 0 \tag{1.78}$$

when D is constant. Further, if the boundary is straight, (1.78) reduces to

$$\frac{\partial^3 w}{\partial n^3} + (2-\nu)\frac{\partial^3 w}{\partial n\,\partial s^2} = 0. \tag{1.79}$$

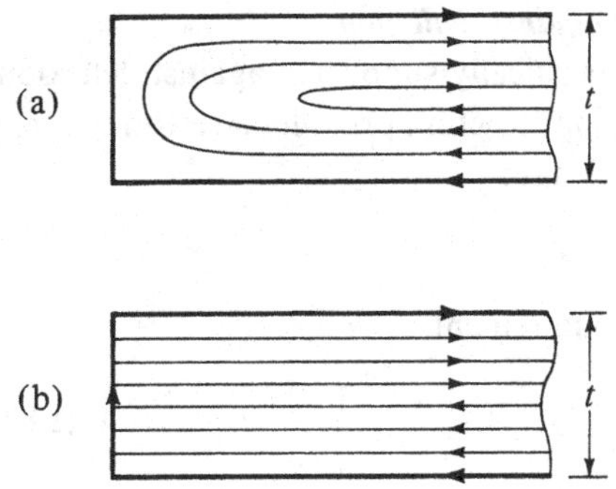

Fig. 1.9

Error involved at a free boundary. An estimate can be made of the error involved in the use of the joint requirement (1.75) – and hence in assumption (i) of the classical theory of plates – by a comparison with an exact three-dimensional solution of a plate problem. Such a solution is provided by the torsion of a long strip of rectangular section. Away from the edges there is agreement with classical plate theory, but in the neighbourhood of the edges the true shear flow is as shown in Fig. 1.9a which may be compared with that of classical plate theory shown in Fig. 1.9b. It is seen that the region of disagreement is limited to a distance of about $1.5t$ from the edges. For a more elaborate theory of plates, which takes account of the shear distortion of the plate, the reader is referred to the work of Reissner (1947).

Plate thickness tapering to zero. If the plate thickness tapers to zero so that in the neighbourhood of a boundary

$$t \sim \omega \boldsymbol{n} + \mathrm{O}(\boldsymbol{n}^2), \tag{1.80}$$

where ω is the (small) taper angle, the rigidity varies as

$$D \sim \frac{E\omega^3 \boldsymbol{n}^3}{12(1-\nu^2)} + \mathrm{O}(\boldsymbol{n}^4). \tag{1.81}$$

It follows that at the boundary

$$D = \frac{\partial D}{\partial s} = \frac{\partial D}{\partial \boldsymbol{n}} = 0, \tag{1.82}$$

and hence (1.76) and (1.77) are satisfied by any smoothly varying deflexion function.

1.7.5 *Edge elastically supported against deflexion*

If an edge, otherwise unloaded, is elastically supported against deflexion in such a manner that the surrounding structure provides a restoring force per unit length equal to ϱw, say, we have

$$M_n = 0,$$

together with the following joint requirement

$$Q_n + \frac{\partial M_{ns}}{\partial s} + \varrho w = 0. \tag{1.83}$$

Equation (1.83) can be expressed in terms of w and its derivatives in the same way as in Section 1.7.4. In particular, if D is constant and the edge is straight, we find

$$\frac{\partial^3 w}{\partial n^3} + (2 - \nu)\frac{\partial^3 w}{\partial n\, \partial s^2} - \frac{\varrho w}{D} = 0. \tag{1.84}$$

1.7.6 *Temperature gradient through the thickness*

If there is a temperature gradient through the thickness, the boundary conditions which involve moments and shears are modified because of the presence of the term κ_T in the moment–curvature relations (1.62). For example, the vanishing of the moment M_n at a simply supported boundary requires

$$\frac{\partial^2 w}{\partial n^2} + \nu \frac{\partial \psi}{\partial s}\frac{\partial w}{\partial n} - (1 + \nu)\kappa_T = 0. \tag{1.85}$$

1.8 Anisotropic plates

Anisotropic materials such as wood have been used as load-bearing members for thousands of years. More recently, high-strength fibres of glass or carbon, for example, have been used in a bonding matrix to make structural components with particular design characteristics. In the following we consider the small-deflexion behaviour of anisotropic plates, particularly multi-layered plates, drawing heavily on the work of Lekhnitsky (1941), E. Reissner and Stavsky (1961) and Stavsky (1961). We note first that the concept of a neutral surface has, in general, no part to play in the analysis of multi-layered plates because of coupling between moments and planar strains.

1.8.1 *Coupled stress–strain and moment–curvature relations*

Consider an element of a multi-layered plate bounded by surfaces at $z = \pm\frac{1}{2}t$ and subjected to constant moments and middle-surface forces. The planar strains at any point in the plate are conveniently expressed in terms of the mid-surface strains

$$\boldsymbol{\varepsilon}^0 = (\varepsilon_x^0, \varepsilon_y^0, \varepsilon_{xy}^0)^{\mathrm{T}}, \tag{1.86}$$

where the affix 0 refers to the plane $z = 0$, and the curvatures

$$\boldsymbol{\kappa} = (\kappa_x, \kappa_y, 2\kappa_{xy})^{\mathrm{T}}, \tag{1.87}$$

where the factor 2, introduced to compensate for our use of the 'engineering' definition of shear strain, enables us to express the strains in a plane at a distance z from the mid-surface in the following simple form, see (1.2),

$$\boldsymbol{\varepsilon} = (\varepsilon_x, \varepsilon_y, \varepsilon_{xy})^{\mathrm{T}} = \boldsymbol{\varepsilon}^0 + z\boldsymbol{\kappa}. \tag{1.88}$$

Now the most general form of plane stress anisotropy is such that

$$\left.\begin{aligned} (\sigma_x, \sigma_y, \tau_{xy})^{\mathrm{T}} &= \boldsymbol{\sigma}, \text{ say} \\ &= \mathbf{E}\boldsymbol{\varepsilon}, \end{aligned}\right\} \tag{1.89}$$

where $\mathbf{E}$ is a symmetrical 3×3 elastic moduli matrix of the form

$$\mathbf{E} = (E_{ij}), \quad i, j = 1, 2, 6, \tag{1.90}$$

where the E_{ij} vary from layer to layer so that $\mathbf{E}$ is a function of z.

The stresses $\boldsymbol{\sigma}$ may be integrated through the thickness of the plate to yield the following resultant middle-surface forces $\mathbf{N}$ and moments $\mathbf{M}$ per unit length, where

$$\left.\begin{aligned} \mathbf{N} &= (N_x, N_y, N_{xy})^{\mathrm{T}} \\ &= \int_{-\frac{1}{2}t}^{\frac{1}{2}t} \boldsymbol{\sigma}\,\mathrm{d}z \\ &= \int_{-\frac{1}{2}t}^{\frac{1}{2}t} \mathbf{E}(\boldsymbol{\varepsilon}^0 + z\boldsymbol{\kappa})\,\mathrm{d}z, \end{aligned}\right\} \tag{1.91}$$

and

$$\left.\begin{aligned} \mathbf{M} &= (M_x, M_y, M_{xy})^{\mathrm{T}} \\ &= \int_{-\frac{1}{2}t}^{\frac{1}{2}t} z\boldsymbol{\sigma}\,\mathrm{d}z \\ &= \int_{-\frac{1}{2}t}^{\frac{1}{2}t} z\mathbf{E}(\boldsymbol{\varepsilon}^0 + z\boldsymbol{\kappa})\,\mathrm{d}z. \end{aligned}\right\} \tag{1.92}$$

At this point it is convenient to introduce the following matrices of 'elastic areas', 'elastic moments of area' and 'elastic second moments of area':

$$\begin{aligned} \mathbf{A} &= (A_{ij}) \\ &= \int_{-\frac{1}{2}t}^{\frac{1}{2}t} \mathbf{E}\,\mathrm{d}z, \end{aligned}$$

$$\left.\begin{aligned} \mathbf{B} &= (B_{ij}) \\ &= \int_{-\frac{1}{2}t}^{\frac{1}{2}t} z\mathbf{E}\,dz \\ \mathbf{D} &= (D_{ij}) \\ &= \int_{-\frac{1}{2}t}^{\frac{1}{2}t} z^2\mathbf{E}\,dz. \end{aligned}\right\} \tag{1.93}$$

Equations (1.91)–(1.93) may now be written concisely as

$$\begin{bmatrix} \mathbf{N} \\ \mathbf{M} \end{bmatrix} = \begin{bmatrix} \mathbf{A} & \mathbf{B} \\ \mathbf{B} & \mathbf{D} \end{bmatrix} \begin{bmatrix} \boldsymbol{\varepsilon}^0 \\ \boldsymbol{\kappa} \end{bmatrix}, \tag{1.94}$$

which it is often convenient to express in the following partly inverted form:

$$\begin{bmatrix} \boldsymbol{\varepsilon}^0 \\ \mathbf{M} \end{bmatrix} = \begin{bmatrix} \mathbf{a} & \mathbf{b} \\ -\mathbf{b}^{\mathrm{T}} & \mathbf{d} \end{bmatrix} \begin{bmatrix} \mathbf{N} \\ \boldsymbol{\kappa} \end{bmatrix}, \tag{1.95}$$

where the affix T denotes the transposition and

$$\left.\begin{aligned} \mathbf{a} &= (a_{ij}) = \mathbf{A}^{-1}, \\ \mathbf{b} &= (b_{ij}) = -\mathbf{A}^{-1}\mathbf{B}, \\ \mathbf{d} &= (d_{ij}) = \mathbf{D} - \mathbf{B}\mathbf{A}^{-1}\mathbf{B}. \end{aligned}\right\} \tag{1.96}$$

Note that **a** and **d** are symmetrical matrices whereas **b** is not.

1.8.2 Equilibrium and compatibility

The normal equilibrium of an element of plate is given by (1.27), where, to account for forces in the plane of the plate, q is replaced by $(q+q')$, where q' is given by (1.40a). Equilibrium in the plane of the plate requires that (1.32) be satisfied and this is achieved by introducing the force function $\boldsymbol{\Phi}$ in (1.33). The condition of compatibility requires that (1.35) be satisfied.

In terms of the deflexion w and force function Φ, (1.95) and the equation of normal equilibrium can be expressed in the form

$$\mathrm{L}_1 w + \mathrm{L}_3 \Phi = q + \lozenge^4(\Phi, w), \tag{1.97}$$

and the condition of compatibility in the form

$$\mathrm{L}_2 \Phi - \mathrm{L}_3 w = 0, \tag{1.98}$$

where

$$\mathrm{L}_1 = \{d_{11}, 4d_{16}, 2(d_{12}+2d_{66}), 4d_{26}, d_{22}\}\boldsymbol{\Delta},$$
$$\mathrm{L}_2 = \{a_{22}, -2a_{26}, 2a_{12}+a_{66}, -2a_{16}, a_{11}\}\boldsymbol{\Delta},$$
$$\mathrm{L}_3 = \{b_{21}, 2b_{26}-b_{61}, b_{11}+b_{22}-2b_{66}, 2b_{16}-b_{62}, b_{12}\}\boldsymbol{\Delta},$$

and $\boldsymbol{\Delta}$ is a column vector of differential operators

$$\boldsymbol{\Delta}=\left(\frac{\partial^4}{\partial x^4},\frac{\partial^4}{\partial x^3\partial y},\frac{\partial^4}{\partial x^2\partial y^2},\frac{\partial^4}{\partial x\partial y^3},\frac{\partial^4}{\partial y^4}\right)^{\mathrm{T}}.$$

1.8.3 Zero coupling between N *and* M

This important class of anisotropy occurs if the operator L_3, and hence **B**, is zero, as in a composite plate with a symmetrical lay-up of fibres, that is, one in which

$$[\mathbf{E}]_{+z}=[\mathbf{E}]_{-z}. \tag{1.99}$$

For such plates the small-deflexion equations assume the form

$$\mathrm{L}_1 w = q + \lozenge^4(\Phi, w), \tag{1.100}$$

and

$$\mathrm{L}_2\Phi = 0. \tag{1.101}$$

Furthermore, if no middle-surface forces are applied to the boundary, Φ is zero throughout the plate and (1.100) assumes the simple form

$$\mathrm{L}_1 w = q. \tag{1.102}$$

Also, because of the vanishing of **B** the coefficients d_{ij} in L_1 are given simply by

$$d_{ij}=D_{ij}.$$

Orthotropic plates. In many practical applications the plate has orthotropic properties aligned to the x, y-axes, so that

$$A_{16}=A_{26}=D_{16}=D_{26}=0.$$

Equation (1.100) then assumes the simple form

$$\left.\begin{aligned}&D_{11}\frac{\partial^4 w}{\partial x^4}+2H\frac{\partial^4 w}{\partial x^2\partial y^2}+D_{22}\frac{\partial^4 w}{\partial y^4}=q+\lozenge^4(\Phi,w),\\&\text{where}\\&H=D_{12}+2D_{66},\end{aligned}\right\} \tag{1.103}$$

and (1.101) becomes

$$a_{22}\frac{\partial^4\Phi}{\partial x^4}+(2a_{12}+a_{66})\frac{\partial^4\Phi}{\partial x^2\partial y^2}+a_{11}\frac{\partial^4\Phi}{\partial y^4}=0. \tag{1.104}$$

1.8.4 Plate with antisymmetrical fibre lay-up

If the lay-up of fibres in a plate is such that for positive values of z the fibre orientation is $+\theta_z$, say, while for negative values of z it is $-\theta_z$, we can write

$$[\mathbf{E}]_{\pm z} = \begin{pmatrix} E_{11} & E_{12} & \operatorname{sig}(z)E_{16} \\ E_{12} & E_{22} & \operatorname{sig}(z)E_{26} \\ \operatorname{sig}(z)E_{16} & \operatorname{sig}(z)E_{26} & E_{66} \end{pmatrix}, \tag{1.105}$$

where sig (z) is $+1$ if $z > 0$ and -1 if $z < 0$.

The resulting equations exhibit a coupling between **N** and **M**, but they are markedly simpler than those for the general case because the following terms vanish:

$$A_{16} = A_{26} = B_{11} = B_{12} = B_{22} = B_{66} = D_{16} = D_{26} = 0. \tag{1.106}$$

It follows from (1.96) that

$$a_{16} = a_{26} = b_{11} = b_{12} = b_{21} = b_{22} = b_{66} = d_{16} = d_{26} = 0.$$

and the L-operators are given by

$$\left.\begin{aligned} \mathrm{L}_1 &= d_{11}\frac{\partial^4}{\partial x^4} + 2(d_{12} + 2d_{66})\frac{\partial^4}{\partial x^2 \partial y^2} + d_{22}\frac{\partial^4}{\partial y^4}, \\ \mathrm{L}_2 &= a_{22}\frac{\partial^4}{\partial x^4} + (2a_{12} + a_{66})\frac{\partial^4}{\partial x^2 \partial y^2} + a_{11}\frac{\partial^4}{\partial y^4}, \\ \mathrm{L}_3 &= (2b_{26} - b_{61})\frac{\partial^4}{\partial x^3 \partial y} + (2b_{16} - b_{62})\frac{\partial^4}{\partial x \partial y^3}. \end{aligned}\right\} \tag{1.107}$$

1.8.5 Plate with unsymmetrical cross-plies

Another class of plates which yield relatively simple equations is one in which for positive z the fibres are aligned with the x-axis while for negative z the fibres are aligned with the y-axis. More generally, such plates consist of pairs of similar orthotropic layers equally disposed about the mid-plane, but with their major principal axes at 0° and 90°, respectively (Whitney and Leissa, 1969). For such plates

$$A_{16} = A_{26} = B_{12} = B_{16} = B_{26} = B_{66} = D_{16} = D_{26} = 0. \tag{1.108}$$

and

$$A_{22} = A_{11}, \quad B_{22} = -B_{11}, \quad D_{22} = D_{11}, \tag{1.109}$$

and hence

$$a_{16} = a_{26} = b_{16} = b_{26} = b_{61} = b_{62} = b_{66} = d_{16} = d_{26} = 0,$$

and

$$b_{12} = -b_{21}.$$

1.8.6 General equations in terms of displacements

When there are clamped boundaries to a multi-layered anisotropic plate with coupling between moments and planar strains, there can be advantages in expressing the governing equations in terms of the deflexion w and the displacements u, v in the plane of the middle surface. Following Whitney and Leissa (1969), we first express $\mathbf{N}, \mathbf{M}$ in terms of these displacements via (1.94) and the relations

$$\left.\begin{aligned} \boldsymbol{\varepsilon}^0 &= \left[\frac{\partial u}{\partial x}, \frac{\partial v}{\partial y}, \frac{\partial u}{\partial y} + \frac{\partial v}{\partial x}\right]^{\mathrm{T}}, \\ \boldsymbol{\kappa} &= -\left[\frac{\partial^2 w}{\partial x^2}, \frac{\partial^2 w}{\partial y^2}, 2\frac{\partial^2 w}{\partial x\,\partial y}\right]^{\mathrm{T}}. \end{aligned}\right\} \tag{1.110}$$

Equilibrium in the plane of the plate, see (1.32), now leads to the following equations in terms of the displacements:

$$\begin{aligned} &A_{11}\frac{\partial^2 u}{\partial x^2} + 2A_{16}\frac{\partial^2 u}{\partial x \partial y} + A_{66}\frac{\partial^2 u}{\partial y^2} + A_{16}\frac{\partial^2 v}{\partial x^2} + (A_{12} + A_{66})\frac{\partial^2 v}{\partial x \partial y} \\ &\quad + A_{26}\frac{\partial^2 v}{\partial y^2} - B_{11}\frac{\partial^3 w}{\partial x^3} - 3B_{16}\frac{\partial^3 w}{\partial x^2\,\partial y} \\ &\quad - (B_{12} + 2B_{66})\frac{\partial^3 w}{\partial x \partial y^2} - B_{26}\frac{\partial^3 w}{\partial y^3} = 0, \end{aligned} \tag{1.111}$$

$$\begin{aligned} &A_{16}\frac{\partial^2 u}{\partial x^2} + (A_{12} + A_{66})\frac{\partial^2 u}{\partial x \partial y} + A_{26}\frac{\partial^2 u}{\partial y^2} + A_{66}\frac{\partial^2 v}{\partial x^2} \\ &\quad + 2A_{26}\frac{\partial^2 v}{\partial x \partial y} + A_{22}\frac{\partial^2 v}{\partial y^2} - B_{16}\frac{\partial^3 w}{\partial x^3} - (B_{12} + 2B_{66})\frac{\partial^3 w}{\partial x^2 \partial y} \\ &\quad - 3B_{26}\frac{\partial^3 w}{\partial x \partial y^2} - B_{22}\frac{\partial^3 w}{\partial y^3} = 0. \end{aligned} \tag{1.112}$$

The equation of normal equilibrium is best derived from (1.27) and (1.40), rather than (1.97), that is, from

$$\begin{aligned} &\frac{\partial^2 M_x}{\partial x^2} + 2\frac{\partial^2 M_{xy}}{\partial x \partial y} + \frac{\partial^2 M_y}{\partial y^2} + q + N_x\frac{\partial^2 w}{\partial x^2} \\ &\quad + 2N_{xy}\frac{\partial^2 w}{\partial x\,\partial y} + N_y\frac{\partial^2 w}{\partial y^2} = 0. \end{aligned} \tag{1.113}$$

Thus, we find

$$D_{11}\frac{\partial^4 w}{\partial x^4} + 4D_{16}\frac{\partial^4 w}{\partial x^3 \partial y} + 2(D_{12} + 2D_{66})\frac{\partial^4 w}{\partial x^2 \partial y^2} + 4D_{26}\frac{\partial^4 w}{\partial x \partial y^3}$$

$$+D_{22}\frac{\partial^4 w}{\partial y^4}-B_{11}\frac{\partial^3 u}{\partial x^3}-3B_{16}\frac{\partial^3 u}{\partial x^2\partial y}-(B_{12}+2B_{66})\frac{\partial^3 u}{\partial x\partial y^2}$$

$$-B_{26}\frac{\partial^3 u}{\partial y^3}-B_{16}\frac{\partial^3 v}{\partial x^3}-(B_{12}+2B_{66})\frac{\partial^3 v}{\partial x^2\partial y}$$

$$-3B_{26}\frac{\partial^3 v}{\partial x\partial y^2}-B_{22}\frac{\partial^3 v}{\partial y^3}=q+\Diamond^4(\Phi,w), \tag{1.114}$$

where

$$\Diamond^4(\Phi,w)=\frac{\partial^2 w}{\partial x^2}\left\{A_{11}\frac{\partial u}{\partial x}+A_{12}\frac{\partial v}{\partial y}+A_{16}\left(\frac{\partial u}{\partial y}+\frac{\partial v}{\partial x}\right)\right\}$$

$$+2\frac{\partial^2 w}{\partial x\partial y}\left\{A_{16}\frac{\partial u}{\partial x}+A_{26}\frac{\partial v}{\partial y}+A_{66}\left(\frac{\partial u}{\partial y}+\frac{\partial v}{\partial x}\right)\right\}$$

$$+\frac{\partial^2 w}{\partial y^2}\left\{A_{12}\frac{\partial u}{\partial x}+A_{22}\frac{\partial v}{\partial y}+A_{26}\left(\frac{\partial u}{\partial y}+\frac{\partial v}{\partial x}\right)\right\}$$

$$-B_{11}\left(\frac{\partial^2 w}{\partial x^2}\right)^2-2B_{12}\frac{\partial^2 w\,\partial^2 w}{\partial x^2\partial y^2}-B_{22}\left(\frac{\partial^2 w}{\partial y^2}\right)^2$$

$$-4\frac{\partial^2 w}{\partial x\partial y}\left(B_{16}\frac{\partial^2 w}{\partial x^2}+B_{66}\frac{\partial^2 w}{\partial x\partial y}+B_{26}\frac{\partial^2 w}{\partial y^2}\right). \tag{1.115}$$

Note that, for initially unstressed plates subjected only to a normal loading, the term $\Diamond^4(\Phi,w)$ varies as the square of the deflexion and should therefore be omitted in linear small-deflexion theory.

1.8.7 *Thermal stress effects*

If $T=T(x,y,z)$ is the temperature at any point in a multi-layered plate, we modify (1.89) to read

$$\left.\begin{aligned}&\boldsymbol{\sigma}=\mathbf{E}(\boldsymbol{\varepsilon}-\boldsymbol{\varepsilon}_T),\\ &\text{where}\\ &\boldsymbol{\varepsilon}_T=[\alpha_1 T,\alpha_2 T,\alpha_6 T]^{\mathrm{T}},\end{aligned}\right\} \tag{1.116}$$

and α_i are the respective coefficients of thermal expansion in a layer at distance z from the mid-thickness. It follows that (1.94) now becomes

$$\begin{bmatrix}\mathbf{N}\\ \mathbf{M}\end{bmatrix}=\begin{bmatrix}\mathbf{A} & \mathbf{B}\\ \mathbf{B} & \mathbf{D}\end{bmatrix}\begin{bmatrix}\boldsymbol{\varepsilon}^0\\ \boldsymbol{\kappa}\end{bmatrix}-\begin{bmatrix}\mathbf{N}_T\\ \mathbf{M}_T\end{bmatrix}, \tag{1.117}$$

where

$$\left.\begin{aligned}\mathbf{N}_T&=\int_{-\frac{1}{2}t}^{\frac{1}{2}t}\mathbf{E}\boldsymbol{\varepsilon}_T\,\mathrm{d}z,\\ \mathbf{M}_T&=\int_{-\frac{1}{2}t}^{\frac{1}{2}t}z\mathbf{E}\boldsymbol{\varepsilon}_T\,\mathrm{d}z.\end{aligned}\right\} \tag{1.118}$$

Likewise, (1.95) becomes

$$\begin{bmatrix} \boldsymbol{\varepsilon}^0 \\ \mathbf{M}+\mathbf{M}_T \end{bmatrix} = \begin{bmatrix} a & b \\ -b^{\mathrm{T}} & d \end{bmatrix} \begin{bmatrix} \mathbf{N}+\mathbf{N}_T \\ \boldsymbol{\kappa} \end{bmatrix}. \tag{1.119}$$

Equations (1.117) and (1.119) enable us to express the governing differential equations in terms of w, Φ as in Section 1.8.2, or in terms of the displacements u, v, w, as in Section 1.8.6. Thus, writing

$$\mathbf{N}_T = [N_{x,T}, N_{y,T}, N_{xy,T}]^{\mathrm{T}} \quad \text{and} \quad \mathbf{M}_T = [M_{x,T}, M_{y,T}, M_{xy,T}]^{\mathrm{T}},$$

we find that in terms of w, Φ the equation of normal equilibrium becomes

$$\begin{aligned} &\mathrm{L}_1 w + \mathrm{L}_3 \Phi + \frac{\partial^2}{\partial x^2}(M_{x,T} + b_{11} N_{x,T} + b_{21} N_{y,T} + b_{61} N_{xy,T}) \\ &\quad + \frac{\partial^2}{\partial y^2}(M_{y,T} + b_{12} N_{x,T} + b_{22} N_{y,T} + b_{62} N_{xy,T}) \\ &\quad + 2\frac{\partial^2}{\partial x \partial y}(M_{xy,T} + b_{16} N_{x,T} + b_{26} N_{y,T} + b_{66} N_{xy,T}) \\ &= q + \lozenge^4(\Phi, w), \end{aligned} \tag{1.120}$$

and the *condition of compatibility* is

$$\begin{aligned} &\mathrm{L}_2 \Phi - \mathrm{L}_3 w + \frac{\partial^2}{\partial x^2}(a_{12} N_{x,T} + a_{22} N_{y,T} + a_{26} N_{xy,T}) \\ &\quad + \frac{\partial^2}{\partial y^2}(a_{11} N_{x,T} + a_{12} N_{y,T} + a_{16} N_{xy,T}) \\ &\quad - \frac{\partial^2}{\partial x \, \partial y}(a_{16} N_{x,T} + a_{26} N_{y,T} + a_{66} N_{xy,T}) = 0. \end{aligned} \tag{1.121}$$

1.8.8 Boundary conditions

When there is coupling between moments and planar strains, the specification of the boundary conditions becomes more complex because attention must now be given to the boundary displacements, or lack of displacements, in the plane of the plate. Thus, if at an edge where w is zero a further condition is such that a rotation $\partial w/\partial \boldsymbol{n}$ of the plate is resisted by a moment $\chi \partial w/\partial \boldsymbol{n}$ due to the surrounding structure, we have

$$M_n = \chi \frac{\partial w}{\partial \boldsymbol{n}}. \tag{1.122}$$

Likewise, if a planar displacement u_n normal to the boundary is resisted by a force per unit length $\chi_n u_n$, we have

$$N_n = \chi_n u_n, \tag{1.123}$$

and similarly if a planar displacement u_t tangential to the boundary is resisted by a force per unit length $\chi_t u_t$, we have

$$N_t = \chi_t u_t. \tag{1.124}$$

Note that unless χ_n and χ_t are zero, the boundary conditions necessarily involve the planar displacements and, in the analysis of such cases, it would be appropriate to work in terms of the displacements. This means that in the boundary equations M_n, N_n and N_t would be given by (1.94). Likewise, if χ_n and χ_t are zero, the analysis would be more appropriate in terms of w and Φ, in which case M_n would be given by (1.95). Finally, we note that if χ, χ_n and χ_t are infinite, the boundary conditions are

$$\left.\begin{aligned} w = \frac{\partial w}{\partial n} = 0, \\ u = v = 0, \end{aligned}\right\} \tag{1.125}$$

which can be described as *fully clamped.* The analysis for such a case is given in Section 3.10.

References

Lekhnitsky, S. G. Bending of nonhomogeneous anisotropic thin plate of symmetrical construction. *J. Appl. Maths. Mech.*, PMM 5, pp. 71–97 (1941). [For a more readily accessible version, see *Anisotropic plates*, Gordon and Breach, New York, 1968.]

Marcus, H. *Die Theorie elastischer Gewebe*, 2nd ed. Berlin, 1932.

Morley, L. S. D. Variational reduction of the clamped plate to two successive membrane problems with an application to uniformly loaded sectors. *Quart. J. Mech. Appl. Math.*, **16**, p. 451 (1963).

Reissner, E. On bending of elastic plates. *Q. Appl. Maths.*, **5**, p. 55 (1947).

Reissner, E., and Stavsky, Y. Bending and stretching of certain types of heterogeneous aeolotropic elastic plates. *J. Appl. Mech.*, **28**, pp. 402–8 (1961).

Stavsky, Y. Bending and stretching of laminated aeolotropic plates, *Proc. ASCE*, **87**, EM 6, pp. 31–56 (1961).

Whitney, J. M., and Leissa, A. W. Analysis of heterogeneous anisotropic plates. *J. Appl. Mech.*, **36**, pp. 261–6 (1969).

Additional references

Ashton, J. E. Analysis of anisotropic plates. *J. of Composite Materials*, **3**, pp. 148–65, 470–9 (1969); **4**, pp. 162–71 (1970).

Jones, R. M. *Mechanics of composite materials*, McGraw-Hill, 1975.

Reissner, E. On transverse bending of plates, including the effect of transverse shear deformation, *Int. J. Solids Structures*, **11**, pp. 569–73 (1975).

———. On the transverse bending of elastic plates. *Int. J. Solids Structures*, **12**, pp. 545–54 (1976).

Whitney, J. M., and Pagano, N. J. Shear deformation in heterogeneous anisotropic plates. *J. Appl. Mech.*, **37**, pp. 1031–6 (1970).

2

Rectangular plates

In this chapter attention is given to methods of solution of the small-deflexion equations for rectangular plates of constant thickness. Timoshenko and Woinowsky-Krieger (1959) have presented a large number of detailed solutions to particular problems of this class and it is not the intention here to duplicate this work, although some overlapping is unavoidable, but rather to present the different methods of solution available.

2.1 Plates with all edges simply supported – double Fourier series solution

Consider first the rectangular plate of sides a, b shown in Fig. 2.1 under the action of a distributed loading of the form

$$q(x, y) = \sum_{m=1}^{\infty} \sum_{n=1}^{\infty} q_{mn} \sin\frac{m\pi x}{a} \sin\frac{n\pi y}{b}, \tag{2.1}$$

where q_{mn} are constants and m, n are integers. Such a series can, of course, represent any distribution of applied loading.

In using a Fourier series representation for the deflexion, care must be exercised to ensure that no unjustifiable differentiations of the series are carried out. This difficulty was overcome by Hopkins (1945) by representing $\partial^8 w/\partial x^4 \partial y^4$ as a double Fourier series,

$$\frac{\partial^8 w}{\partial x^4\, \partial y^4} = \sum_{m=1}^{\infty} \sum_{n=1}^{\infty} A_{mn} \left(\frac{m\pi}{a}\right)^3 \left(\frac{n\pi}{b}\right)^3 \sin\frac{m\pi x}{a} \sin\frac{n\pi y}{b} \tag{2.2}$$

in which the factor $(m\pi/a)^3(n\pi/b)^3$ is introduced merely for convenience.

Expression (2.2) may legitimately be integrated, term by term, any number of times, so that we may write

$$w = \sum_{m=1}^{\infty} \sum_{n=1}^{\infty} A_{mn} X_m Y_n,$$

where

$$X_m = a_m + b_m x + c_m x^2 + d_m x^3 - \frac{a}{m\pi} \sin\frac{m\pi x}{a}, \tag{2.3}$$

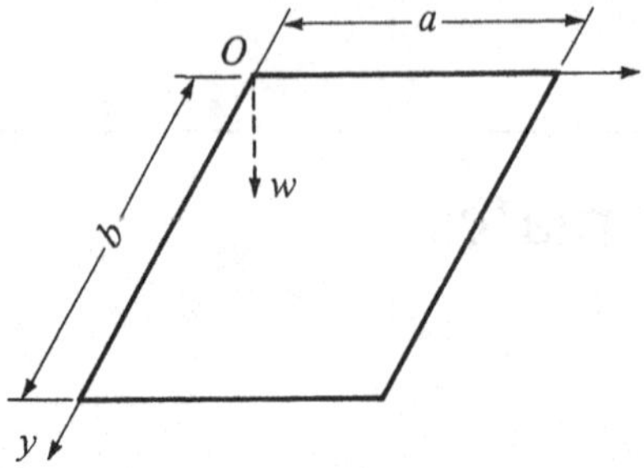

Fig. 2.1

$$Y_n = e_n + f_n y + g_n y^2 + h_n y^3 - \frac{b}{n\pi}\sin\frac{n\pi y}{b}.$$

The four unknown constants appropriate to each integer m and n are sufficient to satisfy the eight boundary conditions. If the edges are simply supported, these conditions are given by (1.52) and (1.54):

$$\left.\begin{array}{ll} [w=0]_{x=0,a}, & \left[\dfrac{\partial^2 w}{\partial x^2}=0\right]_{x=0,a} \\ [w=0]_{y=0,b}, & \left[\dfrac{\partial^2 w}{\partial y^2}=0\right]_{y=0,b}. \end{array}\right\} \tag{2.4}$$

Substitution of (2.3) in (2.4) then yields the simple result

$$\left.\begin{array}{l} a_m = b_m = c_m = d_m = 0 \\ e_n = f_n = g_n = h_n = 0. \end{array}\right\} \tag{2.5}$$

Thus *if the edges are simply supported* the deflexion can be represented by the double Fourier series

$$w = \sum_{m=1}^{\infty}\sum_{n=1}^{\infty} w_{mn}\sin\frac{m\pi x}{a}\sin\frac{n\pi y}{b} \tag{2.6}$$

which can legitimately be differentiated, term by term, up to four times with respect to x and y. Having established this result we may substitute (2.1) and (2.6) in (1.30) and equate coefficients of like terms to give

$$w_{mn} = \frac{q_{mn}}{D\pi^4\left(\dfrac{m^2}{a^2}+\dfrac{n^2}{b^2}\right)^2}. \tag{2.7}$$

This solution was first given by Navier (1820).

Special Cases. If the loading is uniform and equal to q_0, the term q_{mn} is given by

$$q_{mn} = \frac{16q_0}{\pi^2 mn}, \tag{2.8}$$

where m and n are odd integers.

If the loading varies linearly and is given by

$$\left.\begin{aligned} q(x, y) &= q_1 x/a, \\ q_{mn} &= \frac{4q_1}{\pi^2 mn}(-1)^m\{(-1)^n - 1\}. \end{aligned}\right\} \tag{2.9}$$

If there is a concentrated load P at the point $x = a'$, $y = b'$,

$$q_{mn} = \frac{4P}{ab}\sin\frac{m\pi a'}{a}\sin\frac{n\pi b'}{b}. \tag{2.10}$$

2.1.1 *Orthotropic plate*

A similar analysis is possible for an orthotropic plate when substitution of (2.1) and (2.6) into (1.103) yields

$$w_{mn} = \frac{q_{mn}}{\pi^4\left(\dfrac{m^4}{a^4}D_{11} + \dfrac{2m^2n^2}{a^2b^2}H + \dfrac{n^4}{b^4}D_{22}\right)}. \tag{2.11}$$

2.1.2 *Effect of middle-surface forces on the deflexion*

The effect of middle-surface forces per unit length, N_x and N_y, can be determined by substituting (2.6) in (1.42) and equating coefficients of like terms to give

$$w_{mn} = \frac{q_{mn}}{D\pi^4\left(\dfrac{m^2}{a^2} + \dfrac{n^2}{b^2}\right)^2 + \pi^2\left(\dfrac{m^2N_x}{a^2} + \dfrac{n^2N_y}{b^2}\right)}. \tag{2.12}$$

By the same token, we find for an orthotropic plate

$$w_{mn} = \frac{q_{mn}}{\pi^4\left(\dfrac{m^4}{a^4}D_{11} + \dfrac{2m^2n^2}{a^2b^2}H + \dfrac{n^4}{b^4}D_{22}\right) + \pi^2\left(\dfrac{m^2}{a^2}N_x + \dfrac{n^2}{b^2}N_y\right)}. \tag{2.13}$$

An inspection of the denominators of the above expressions and a comparison with (2.7) or (2.11) shows that the effect of middle-surface *tensions* is to reduce each deflexion component w_{mn} by a factor r_{mn}, say, where, for the isotropic plate for example,

$$r_{mn} = \left[1 + \left(\frac{m^2 N_x}{a^2} + \frac{n^2 N_y}{b^2}\right) \Big/ \left\{D\pi^2\left(\frac{m^2}{a^2} + \frac{n^2}{b^2}\right)^2\right\}\right]^{-1}. \tag{2.14}$$

When both N_x and N_y are compressive, that is, negative, it follows from (2.14) that all the deflexion components are increased. When N_x, say, is compressive and N_y is tensile, the deflexion components q_{mn} are increased or reduced, according to the sign of $(m^2 N_x/a^2 + n^2 N_y/b^2)$. Thus whenever there is a middle-surface compression some, or all, of the ratios r_{mn} exceed unity. Suppose now that by choosing various integers m, n we find that the greatest of the ratios r_{mn} is $r_{m'n'}$, say. By increasing the magnitude of the middle-surface forces, the ratio $r_{m'n'}$ will increase (and do so at a greater rate than all other ratios) until it becomes infinite. The deflected shape is then dominated by the term

$$w_{m'n'} \sin\frac{m'\pi x}{a} \sin\frac{n'\pi y}{b} \tag{2.15}$$

because $w_{m'n'}$ is theoretically infinite. In practice, of course, such deflexions cannot exist and, further, they violate the assumptions inherent in small-deflexion theory. However, the results can be interpreted in a practical manner and without violating these assumptions by stating that a finite deflexion proportional to $\sin(m'\pi x/a)\sin(n'\pi y/b)$ is possible without the application of normal loads. This phenomenon is called *buckling* and is discussed in greater detail and from a different standpoint in Section 6.2.1. For a much fuller discussion the reader is referred to monographs by Timoshenko (1936), Cox (1962) and Thompson (1973).

2.1.3 *Effect of an elastic foundation*

The effect on the deflexions of an elastic foundation can readily be determined by substituting (2.6) in (1.44). Equating coefficients of like terms, as in the preceding analysis, gives

$$w_{mn} = \frac{q_{mn}}{D\pi^4\left(\dfrac{m^2}{a^2} + \dfrac{n^2}{b^2}\right)^2 + \pi^2\left(\dfrac{m^2 N_x}{a^2} + \dfrac{n^2 N_y}{b^2}\right) + k}. \tag{2.16}$$

The onset of buckling may again be found, as in the previous section, by equating to zero the denominator in (2.16) and determining the smallest values of N_x, N_y for various integral values of m, n. When k is zero, it can be shown that buckling occurs in a mode $\sin(m'\pi x/a)\sin(n'\pi y/b)$ in which either m' or n' is unity. But if k is sufficiently large this may not be so, for the wavelength of the buckles then depends primarily on k and D rather than on a or b. This can be demonstrated most conveniently for the

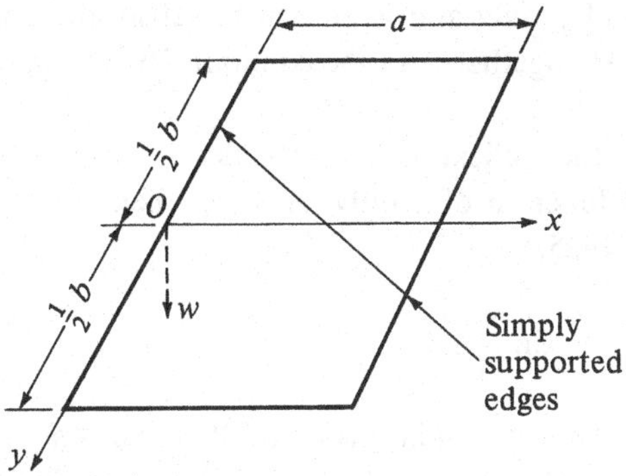

Fig. 2.2

uniformly compressed square plate. Thus, writing $N_x = N_y = -N$, and writing $\mu = a^2/\{\pi^2(m^2 + n^2)\}$, we find from (2.16) that

$$N = k\mu + D/\mu. \tag{2.17}$$

If m or n is a sufficiently large integer we may legitimately regard μ as a continuous, rather than discontinuous, variable; the least value of N may then be found by differentiating (2.17) to give

$$N_{cr} \approx 2\sqrt{Dk} \tag{2.18}$$

which occurs when

$$\mu \approx \sqrt{D/k}.$$

Equation (2.18) is substantially correct if $(m^2 + n^2) \geqslant 32$, say, that is, if

$$k > 10^5 D/a^4.$$

A similar analysis when $N_x = -N$ and $N_y = -\lambda N$ (where $\lambda \leqslant 1$) shows that (2.18) is also applicable. Finally, note that because the wavelength in the direction of N_x is small and virtually independent of the size of the plate, (2.18) is substantially correct for a variety of boundary conditions and shapes if we interpret N_x as the greater of the principal compressive forces/unit length in the plate.

2.2 Plates with two opposite edges simply supported – single Fourier series solution

The use of a single Fourier series solution for plates (Fig. 2.2) with two opposite edges ($x = 0, a$) simply supported was suggested by Lévy (1899). The deflexion is expressed in the form

$$w = \sum_{m=1}^{\infty} Y'_m \sin \frac{m\pi x}{a}, \tag{2.19}$$

where Y'_m is a function of y only and is chosen to satisfy the boundary conditions along $y = \pm\frac{1}{2}b$ together with the equation $D\nabla^4 w = q$, or more generally (1.44).

The form (2.19) for the deflexion is particularly useful when the distributed loading is a function of x only, and it is then convenient to write, following Nadai (1925),

$$w = w_1(x) + \sum_{m=1}^{\infty} Y_m \sin\frac{m\pi x}{a}, \tag{2.20}$$

where w_1 satisfies the boundary conditions at $x = 0, a$ and is a particular integral of the equation $D\nabla^4 w = q$. Indeed, the deflexion w_1 is that of a simply supported 'beam' under the loading $q(x)$ and is determined below in the general case in which

$$\left.\begin{aligned} q = q(x) &= \sum_{m=1}^{\infty} q_m \sin\frac{m\pi x}{a}, \text{ say,} \\ \text{where} \qquad & \\ q_m &= \frac{2}{a}\int_0^a q(x)\sin\frac{m\pi x}{a}\,\mathrm{d}x. \end{aligned}\right\} \tag{2.21}$$

The 'beam' deflexion w_1 is now obtained by integrating the equation

$$D\frac{\mathrm{d}^4 w_1}{\mathrm{d}x^4} = q(x) = \sum_{m=1}^{\infty} q_m \sin\frac{m\pi x}{a} \tag{2.22}$$

subject to the boundary conditions

$$[w_1]_{0,a} = \left[\frac{\mathrm{d}^2 w_1}{\mathrm{d}x^2}\right]_{0,a} = 0. \tag{2.23}$$

The solution of (2.22) and (2.23) is simply

$$w_1 = \frac{a^4}{D\pi^4}\sum_{m=1}^{\infty}\frac{q_m}{m^4}\sin\frac{m\pi x}{a}. \tag{2.24}$$

As for the functions Y_m, they are now chosen to satisfy the boundary conditions at $y = \pm\frac{1}{2}b$ together with the equation

$$\nabla^4\left\{Y_m \sin\frac{m\pi x}{a}\right\} = 0$$

which, on differentiation and division throughout by $\sin(m\pi x/a)$, gives

$$\frac{\mathrm{d}^4 Y_m}{\mathrm{d}y^4} - 2\left(\frac{m\pi}{a}\right)^2\frac{\mathrm{d}^2 Y_m}{\mathrm{d}y^2} + \left(\frac{m\pi}{a}\right)^4 Y_m = 0. \tag{2.25}$$

Equation (2.25) may be integrated to give

$$Y_m = A_m \sinh\frac{m\pi y}{a} + B_m \cosh\frac{m\pi y}{a} + \frac{m\pi y}{a}\left(C_m \sinh\frac{m\pi y}{a} + D_m \cosh\frac{m\pi y}{a}\right). \tag{2.26}$$

The four constants A_m, B_m, C_m, D_m are to be determined from the conditions along $y = \pm\frac{1}{2}b$. If the boundary conditions there are the same, the deflexion will be symmetrical in y, and $A_m = D_m = 0$. Further, if the edges $y = \pm\frac{1}{2}b$ are clamped and we write $m\pi b/2a = \alpha_m$,

$$\begin{aligned}[w]_{y=\pm\frac{1}{2}b} &= \left[w_1 + \sum_{m=1}^{\infty} Y_m \sin\frac{m\pi x}{a}\right]_{y=\pm\frac{1}{2}b} \\ &= \sum_{m=1}^{\infty}\left(\frac{a^4 q_m}{D\pi^4 m^4} + B_m \cosh\alpha_m + \alpha_m C_m \sinh\alpha_m\right)\sin\frac{m\pi x}{a} \\ &= 0 \end{aligned} \tag{2.27}$$

in virtue of (2.20) and (2.24).

Similarly

$$\left[\frac{\partial w}{\partial y}\right]_{y=\pm\frac{1}{2}b} = \sum_{m=1}^{\infty}\frac{2\alpha_m}{b}(B_m \sinh\alpha_m + C_m(\sinh\alpha_m + \alpha_m\cosh\alpha_m))\sin\frac{m\pi x}{a} = 0. \tag{2.28}$$

Equations (2.27) and (2.28) are satisfied by taking

$$\left.\begin{aligned} B_m &= \frac{-a^4 q_m(1+\alpha_m\coth\alpha_m)}{D\pi^4 m^4(\cosh\alpha_m + \alpha_m \operatorname{cosech}\alpha_m)}, \\ C_m &= \frac{a^4 q_m}{D\pi^4 m^4(\cosh\alpha_m + \alpha_m \operatorname{cosech}\alpha_m)}. \end{aligned}\right\} \tag{2.29}$$

A similar analysis for simply supported edges gives

$$\left.\begin{aligned} B_m &= -\frac{a^4 q_m(2+\alpha_m\tanh\alpha_m)}{2D\pi^4 m^4\cosh\alpha_m}, \\ C_m &= \frac{a^4 q_m}{2D\pi^4 m^4\cosh\alpha_m}. \end{aligned}\right\} \tag{2.30}$$

The coefficients B_m and C_m decrease very rapidly as m increases – owing in part to the term m^4 in the denominators of (2.29) and (2.30) – and a satisfactory answer for the deflexion can be obtained by taking only a few terms in the series.

As for the coefficients q_m, it is worth noting that for a uniform loading, q_0,

$$q_m = \frac{2q_0}{\pi m}\{1-(-1)^m\};$$

for a linearly varying loading, $q = q_1 x/a$,

$$q_m = \frac{2q_1}{\pi m}(-1)^{m+1};$$

and if the loading is a 'line load' along $x = l$, of intensity L per unit length,

$$q_m = \frac{2L}{a}\sin\frac{m\pi l}{a}.$$

2.2.1 *Loading and rigidity varying abruptly*

If the loading and/or rigidity is of a discontinuous form given by

$$\left.\begin{aligned} &\left.\begin{aligned} q &= q'(x) \\ D &= D' \end{aligned}\right\} \quad \text{in the range } 0 < y < c, \\ &\left.\begin{aligned} q &= q''(x) \\ D &= D'' \end{aligned}\right\} \quad \text{in the range } c < y < b, \end{aligned}\right\} \tag{2.31}$$

it is possible to apply the method of the previous section separately in each range. There will then be eight sets of coefficients $A'_m, B'_m, \ldots, D''_m$ to be determined from the edge conditions at $y = 0$, b together with the four conditions of continuity at $y = c$. If we write w' and w'' for the deflexion in each range, the conditions at $y = c$ are

$$\left.\begin{aligned} &w' = w'', \\ &\frac{\partial w'}{\partial y} = \frac{\partial w''}{\partial y}, \\ &D'\left(\frac{\partial^2 w'}{\partial y^2} + \nu\frac{\partial^2 w'}{\partial x^2}\right) = D''\left(\frac{\partial^2 w''}{\partial y^2} + \nu\frac{\partial^2 w''}{\partial x^2}\right), \\ &D'\left(\frac{\partial^3 w'}{\partial y^3} + (2-\nu)\frac{\partial^3 w'}{\partial x^2\,\partial y}\right) = D''\left(\frac{\partial^3 w''}{\partial y^3} + (2-\nu)\frac{\partial^3 w''}{\partial x^2\,\partial y}\right), \end{aligned}\right\} \tag{2.32}$$

where the last equation of (2.32) represents the joint requirement of continuity of shear and twisting moment discussed in Section 1.6.5.

2.2.2 *Generalizations of Lévy's method*

The success of the method of Section 2.2 depends on the fact that when $q = q(x)$ it is easy to obtain a particular integral of the equation (1.30) which satisfies the conditions of simple support along the edges $x = 0$, a. There are further load distribution forms for which this is possible, and two such are considered below. No examples are given because once the form of w_1 is determined the method of solution is identical with that of Section 2.2.

Load distribution of the form $q = (y/b)q(x)$. If the load distribution is of the form

$$q = \frac{y}{b} \sum_{m=1}^{\infty} q_m \sin \frac{m\pi x}{a}, \tag{2.33}$$

it may be verified that the deflexion given by

$$w_1 = \frac{ya^4}{bD\pi^4} \sum_{m=1}^{\infty} \frac{q_m}{m^4} \sin \frac{m\pi x}{a} \tag{2.34}$$

satisfies the equation $D\nabla^4 w_1 = q$, together with the conditions of simple support along the edges $x = 0$, a.

Load distribution of the form $q = e^{\beta\pi y/a} q(x)$. If the load distribution is of the form

$$q = e^{\beta\pi y/a} \sum_{m=1}^{\infty} q_m \sin \frac{m\pi x}{a}, \tag{2.35}$$

where β is a constant and the factor π/a has been introduced for convenience, it may be verified that the deflexion given by

$$w_1 = \frac{a^4 e^{\beta\pi y/a}}{D\pi^4} \sum_{m=1}^{\infty} \frac{q_m}{(m^2 - \beta^2)^2} \sin \frac{m\pi x}{a} \tag{2.36}$$

satisfies the equation $D\nabla^4 w_1 = q$, together with the conditions of simple support along the edges $x = 0$, a.

2.2.3 *Effect of middle-surface forces*

In a rectangular plate with two opposite edges simply supported the effect of middle-surface forces per unit length, N_x and N_y, may be determined by the method of Section 2.2.

The deflexion is again represented by the form

$$w = w_1(x) + \sum_{m=1}^{\infty} Y_m \sin \frac{m\pi x}{a}, \tag{2.37}$$

but w_1 is now obtained by integrating the equation

$$D \frac{d^4 w_1}{dx^4} - N_x \frac{d^2 w_1}{dx^2} = \sum_{m=1}^{\infty} q_m \sin \frac{m\pi x}{a}, \tag{2.38}$$

where q_m is given by (2.21).

The solution of (2.38) subject to conditions of simple support along $x=0$, a, is given by

$$w_1 = \sum_{m=1}^{\infty} \frac{q_m \sin \dfrac{m\pi x}{a}}{D\left(\dfrac{m\pi}{a}\right)^4 + N_x\left(\dfrac{m\pi}{a}\right)^2}. \tag{2.39}$$

The functions Y_m now satisfy the differential equation

$$\frac{\mathrm{d}^4 Y_m}{\mathrm{d}y^4} - 2\varphi_1 \frac{\mathrm{d}^2 Y_m}{\mathrm{d}y^2} + \varphi_2 Y_m = 0,$$

where

$$\left.\begin{aligned} \varphi_1 &= \left(\frac{m\pi}{a}\right)^2 + \frac{N_y}{2D}, \\ \varphi_2 &= \left(\frac{m\pi}{a}\right)^2 \left\{\left(\frac{m\pi}{a}\right)^2 + \frac{N_x}{D}\right\}. \end{aligned}\right\} \tag{2.40}$$

The integration of (2.40) assumes one of five forms, depending on the sign and relative magnitudes of φ_1 and φ_2.

Thus, if φ_2 is negative,

$$\left.\begin{aligned} Y_m &= A_m \sinh \alpha_m y + B_m \cosh \alpha_m y \\ &\quad + C_m \sin \beta_m y + D_m \cos \beta_m y, \\ \text{where}& \\ \alpha_m &= \{(\varphi_1^2 - \varphi_2)^{\frac{1}{2}} + \varphi_1\}^{\frac{1}{2}}, \\ \beta_m &= \{(\varphi_1^2 - \varphi_2)^{\frac{1}{2}} - \varphi_1\}^{\frac{1}{2}}. \end{aligned}\right\} \tag{2.41}$$

If $0 < \varphi_2 < \varphi_1^2$ and φ_1 is positive, then

$$\left.\begin{aligned} Y_m &= A_m \sinh \alpha_m y + B_m \cosh \alpha_m y \\ &\quad + C_m \sinh \beta_m y + D_m \cosh \beta_m y, \\ \text{where}& \\ \alpha_m &= \{\varphi_1 + (\varphi_1^2 - \varphi_2)^{\frac{1}{2}}\}^{\frac{1}{2}}, \\ \beta_m &= \{\varphi_1 - (\varphi_1^2 - \varphi_2)^{\frac{1}{2}}\}^{\frac{1}{2}}. \end{aligned}\right\} \tag{2.42}$$

If $0 < \varphi_2 < \varphi_1^2$ and φ_1 is negative, then

$$\left.\begin{aligned} Y_m &= A_m \sin \alpha_m y + B_m \cos \alpha_m y \\ &\quad + C_m \sin \beta_m y + D_m \cos \beta_m y, \\ \text{where}& \\ \alpha_m &= \{-\varphi_1 + (\varphi_1^2 - \varphi_2)^{\frac{1}{2}}\}^{\frac{1}{2}}, \\ \beta_m &= \{-\varphi_1 - (\varphi_1^2 - \varphi_2)^{\frac{1}{2}}\}^{\frac{1}{2}}. \end{aligned}\right\} \tag{2.43}$$

Similarly, if $\varphi_2 > \varphi_1^2$, then

$$\left.\begin{aligned} Y_m &= (A_m \sinh \alpha_m y + B_m \cosh \alpha_m y) \sin \beta_m y \\ &\quad + (C_m \sinh \alpha_m y + D_m \cosh \alpha_m y) \cos \beta_m y, \\ \text{where} \quad & \\ \alpha_m &= \{\tfrac{1}{2}(\varphi_2^{\frac{1}{2}} + \varphi_1)\}^{\frac{1}{2}}, \\ \beta_m &= \{\tfrac{1}{2}(\varphi_2^{\frac{1}{2}} - \varphi_1)\}^{\frac{1}{2}}. \end{aligned}\right\} \tag{2.44}$$

Finally, if $\varphi_2 = \varphi_1^2$, then

$$\begin{aligned} Y_m &= (A_m \sinh \varphi_1^{\frac{1}{2}} y + B_m \cosh \varphi_1^{\frac{1}{2}} y) \\ &\quad + y(C_m \sinh \varphi_1^{\frac{1}{2}} y + D_m \cosh \varphi_1^{\frac{1}{2}} y). \end{aligned} \tag{2.45}$$

2.2.4 *Orthotropic plate*

Similar analyses are possible for an orthotropic plate subjected to a distributed loading which is a function of x only; middle-surface forces N_x, N_y may also be present. For such a plate, (2.20) and (2.21) retain their validity while, from (1.104), equations (2.38) and (2.39) retain their validity with D replaced by D_{11}. Likewise the functions Y_m now satisfy the differential equation

$$\left.\begin{aligned} &\frac{d^4 Y_m}{dy^4} - 2\varphi_1 \frac{d^2 Y_m}{dy^2} + \varphi_2 Y_m = 0, \\ &\text{where } \varphi_1 \text{ and } \varphi_2 \text{ are redefined as} \\ &\varphi_1 = \frac{H}{D_{22}}\left(\frac{m\pi}{a}\right)^2 + \frac{N_y}{2D_{22}}, \\ &\varphi_2 = \frac{D_{11}}{D_{22}}\left(\frac{m\pi}{a}\right)^2 \left\{\left(\frac{m\pi}{a}\right)^2 + \frac{N_x}{D_{11}}\right\}. \end{aligned}\right\} \tag{2.46}$$

The Y_m are thus again given by (2.41)–(2.45).

Further load distributions

In parallel with Section 2.2.2, we find that for the load distribution of (2.33) the function w_1 is given by (2.34), with D replaced by D_{11}. Likewise for the load distribution of (2.35), we find

$$w_1 = \frac{a^4 e^{\beta\pi y/a}}{\pi^2} \sum_{m=1}^{\infty} \frac{q_m \sin(m\pi x/a)}{\pi^2(D_{11}m^4 - 2H\beta^2 m^2 + D_{22}\beta^4) + a^2(N_x - \beta^2 N_y)}. \tag{2.47}$$

2.3 Plates with two opposite edges clamped

The comparative ease with which the deflexion may be determined in plates with an opposite pair of edges simply supported can be attributed to the fact that elementary components of the deflexion exist, each of which – for loadings of the form $q = q(x)$ – satisfies these boundary conditions and the governing differential equation. Morley (1963) derived corresponding components for the deflexion when an opposite pair of edges is clamped. These are more complex than the simple *sin* term in (2.19) and the satisfaction of the other boundary conditions is achieved by a variational procedure (see, for example, Morley 1963, 1964). The method is demonstrated below for the case of a uniformly loaded rectangular plate with all edges clamped.

2.3.1 Plate with all edges clamped

Following Morley (1963), we consider the plate dimensions to be $2a \times 2b$ with the origin at the centre. Under a uniform loading q_0, equation (1.29) may now be integrated to give

$$w = \frac{(b^2 - y^2)^2 q_0}{24D} + w_c, \tag{2.48}$$

where the first term is a particular integral that satisfies the boundary conditions along $y = \pm b$. It follows that the complementary function w_c must satisfy

$$w_c = 0, \quad \partial w_c/\partial y = 0 \quad \text{at } y = \pm b, \tag{2.49}$$

and

$$w_c = -\frac{(b^2 - y^2)^2 q_0}{24D}, \quad \partial w_c/\partial x = 0 \quad \text{at } x = \pm a. \tag{2.50}$$

We now express w_c in the form

$$w_c = \sum_r A_r w_r(x, y), \tag{2.51}$$

where

$$w_r(x, y) = \frac{q_0}{D}(y \sin \lambda_r y \cos \lambda_r b - b \sin \lambda_r b \cos \lambda_r y) \cosh \lambda_r x. \tag{2.52}$$

Each of the functions $w_r(x, y)$ thus satisfies $\nabla^4 w_r = 0$ and vanishes along $y = \pm b$; see (2.49). Further, by choosing the (complex) constants λ_r to be the roots of the transcendental equation

$$2\lambda_r b + \sin 2\lambda_r b = 0, \tag{2.53}$$

the remaining boundary condition along $y = \pm b$ is satisfied. The (complex) coefficients A_r in (2.51) are chosen to satisfy the boundary conditions (2.50) by use of the principle of least work which, for plates with all edges clamped, requires that

$$\int_{-b}^{b}\int_{-a}^{a} \nabla^2 w \nabla^2 \delta w \, dx\, dy = 0, \tag{2.54}$$

which, from (2.48), may be expressed as

$$\int_{-b}^{b}\int_{-a}^{a} \nabla^2 w_c \nabla^2 \delta w_c \, dx\, dy$$

$$= 2\int_{-b}^{b}\left[\frac{(b^2-y^2)^2 q_0}{24D}\frac{\partial}{\partial x}\nabla^2 \delta w_c\right]_{x=a} dy. \tag{2.55}$$

Note that because the method of analysis is based on energy principles, it would have been equally appropriate to present this solution in Chapter 6, where such methods are discussed in greater detail. As for the satisfaction of (2.55), we first note that

$$\nabla^2 w = \frac{2q_0}{D}\sum_r A_r \lambda_r \cos\lambda_r b \cos\lambda_r y \cosh\lambda_r x \tag{2.56}$$

and, for convenience, use the convention

$$\lambda_{-r} = \bar{\lambda}_r; \quad w_{-r}(x, y) = \overline{w_r(x, y)}; \quad \text{etc.}, \tag{2.57}$$

where the bar denotes that the conjugate complex value is to be taken. Thus, w and $\nabla^2 w$ are real quantities when the summations are taken over the positive and negative values of r. As for the term $r=0$, it may be shown that to satisfy the boundary conditions (2.49) it is necessary that $\lambda_0 = 0$.

We now introduce the notation

$$V(w_r, w_s) = V(w_s, w_r) = \int_{-b}^{b}\int_{-a}^{a} \nabla^2 w_r \nabla^2 w_s \, dx\, dy, \tag{2.58}$$

$$V(w_r, w) = 2\int_{-b}^{b}\left[\frac{(b^2-y^2)^2 q_0}{24D}\frac{\partial}{\partial x}\nabla^2 w_r\right]_{x=a} dy. \tag{2.59}$$

Substitution of (2.56) into (2.58), (2.59) yields

$$V(w_r, w_s) = \frac{16 q_0^2 b}{D^2}\frac{\lambda_r \lambda_s}{(\lambda_r^2 - \lambda_s^2)^2}(\lambda_r^2 \cos^2\lambda_s b - \lambda_s^2 \cos^2 \lambda_r b)$$

$$\times (\lambda_s \sinh\lambda_s a \cosh\lambda_r a - \lambda_r \sinh\lambda_r a \cosh\lambda_s a) \tag{2.60}$$

except when $r = s$ and then

$$V(w_r, w_r) = 0; \tag{2.61}$$

it should be noted furthermore that

$$V(w_r, \bar{w}_s) = \overline{V(\bar{w}_r, w_s)} \tag{2.62}$$

and that $V(w_r, \bar{w}_r) \neq 0$, the actual value being obtained by substituting $\lambda_s = \bar{\lambda}_r$ in (2.60). Finally, we have

$$V(w_r, w) = -\frac{8q_0^2 b}{3D^2\lambda_r^2}\{3(1+\cos^2\lambda_r b) - b^2\lambda_r^2\}\sinh\lambda_r a. \tag{2.63}$$

In a practical calculation, the infinite series (2.51) is terminated after the first few terms, say, when $-n \leqslant r \leqslant n$. Equations (2.51) to (2.63) then provide the following n complex simultaneous equations for the determination of the n complex coefficients A_r,

$$\begin{bmatrix} V(w_1,\bar{w}_1) & V(w_1,\bar{w}_2) & \cdots & V(w_1,\bar{w}_n) \\ V(w_2,\bar{w}_1) & V(w_2,\bar{w}_2) & \cdots & V(w_2,\bar{w}_n) \\ \cdots & \cdots & \cdots & \cdots \\ V(w_n,\bar{w}_1) & V(w_n,\bar{w}_2) & \cdots & V(w_n,\bar{w}_n) \end{bmatrix} \begin{bmatrix} \bar{A}_1 \\ \bar{A}_2 \\ \cdot\,\cdot \\ \bar{A}_n \end{bmatrix} + \begin{bmatrix} 0 & V(w_1,w_2) & \cdots & V(w_1,w_n) \\ V(w_2,w_1) & 0 & \cdots & V(w_2,w_n) \\ \cdots & \cdots & \cdots & \cdots \\ V(w_n,w_1) & V(w_n,w_2) & \cdots & 0 \end{bmatrix} \begin{bmatrix} A_1 \\ A_2 \\ \\ A_n \end{bmatrix} = \begin{bmatrix} V(w_1,w) \\ V(w_2,w) \\ \cdot\;\cdot\;\cdot \\ V(w_n,w) \end{bmatrix}. \tag{2.64}$$

Deflexion components for odd functions of y

In the example just given, the deflexion was an even function of y because the loading was uniform. For more general load distributions it may be necessary to include odd components in the expansion (2.51) for w_c. Such components are given by

$$w_r(x, y) = \frac{q_0}{D}(y\cos\lambda_r y\sin\lambda_r b - b\cos\lambda_r b\sin\lambda_r y)\sinh\lambda_r x, \tag{2.65}$$

and the boundary conditions (2.49) are satisfied by choosing the complex constants λ_r to be the roots of the transcendental equation

$$2\lambda_r b - \sin 2\lambda_r b = 0. \tag{2.66}$$

An example using odd and even functions is given in Section 3.7.

2.4 General loading on plates with all edges clamped

When the load distribution cannot be expressed in the form $q = q(x)$, the solution for the plate with all edges clamped may be determined by a

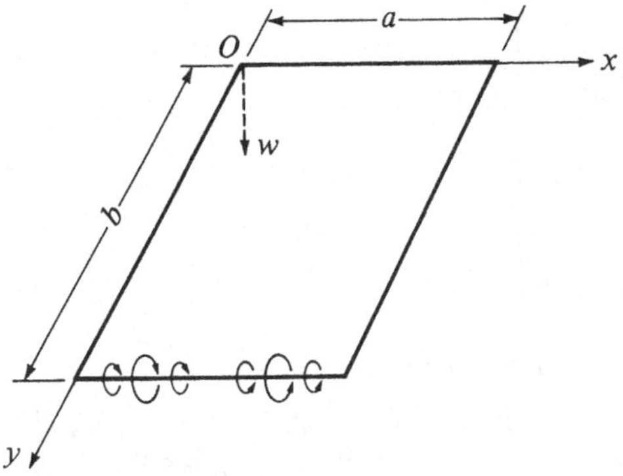

Fig. 2.3

method due to Timoshenko (1938). This method combines the solution for a simply supported plate with that for a plate subjected to moments distributed along the edges. These moments are chosen to satisfy the clamped edge condition or, more generally, conditions of elastic support against rotation.

2.4.1 Simply supported plate with moments applied to one edge

We first determine the deflexion of a rectangular plate due to an arbitrary distribution of moments along the edge $y = b$, the other edges being simply supported (Fig. 2.3). By combining this solution with others obtained by interchanging a and b, and changing the reference axes, it is possible to obtain the deflexion of the plate with an arbitrary distribution of moments around the edges.

As in Section 2.2, we search for a solution in the form

$$w = \sum_{m=1}^{\infty} \sin\frac{m\pi x}{a}\left\{A_m \sinh\frac{m\pi y}{a} + B_m \cosh\frac{m\pi y}{a} + \frac{m\pi y}{a}\left(C_m \sinh\frac{m\pi y}{a} + D_m \cosh\frac{m\pi y}{a}\right)\right\} \tag{2.67}$$

where the constants A_m, B_m, C_m, D_m are to be obtained from the condition of simple support along the edge $y = 0$, while along the edge $y = b$

$$\left.\begin{aligned} &w = 0, \\ &M_{y,b} = -D\frac{\partial^2 w}{\partial y^2} = \sum_{m=1}^{\infty} E_m \sin\frac{m\pi x}{a}, \text{ say.} \end{aligned}\right\} \tag{2.68}$$

Substituting (2.67) in (2.68) and equating coefficients of $\sin m\pi x/a$ gives

$$A_m = \frac{b^2 E_m \cosh\beta_m}{2D\beta_m \sinh^2\beta_m},$$

$$\left.\begin{aligned} &B_m = C_m = 0, \\ &D_m = \frac{-b^2 E_m}{2D\beta_m^2 \sinh \beta_m}, \\ \text{where} \\ &\beta_m = \frac{m\pi b}{a}. \end{aligned}\right\} \tag{2.69}$$

The slopes at the edges are now given by

$$\left.\begin{aligned} \left[\frac{\partial w}{\partial y}\right]_{y=0} &= \frac{b}{2D}\sum_{m=1}^{\infty}\left(\frac{\beta_m \cosh \beta_m - \sinh \beta_m}{\beta_m \sinh^2 \beta_m}\right) E_m \sin\frac{m\pi x}{a}, \\ \left[\frac{\partial w}{\partial y}\right]_{y=b} &= \frac{b}{2D}\sum_{m=1}^{\infty}\left(\frac{\beta_m - \sinh \beta_m \cosh \beta_m}{\beta_m \sinh^2 \beta_m}\right) E_m \sin\frac{m\pi x}{a}, \\ \left[\frac{\partial w}{\partial x}\right]_{x=0} &= \frac{b}{2D}\sum_{m=1}^{\infty}\frac{E_m}{\beta_m \sinh^2 \beta_m}\left(\beta_m \cosh \beta_m \sinh\frac{m\pi y}{a} \right. \\ &\qquad \left. - \sinh \beta_m \cdot \frac{m\pi y}{a}\cosh\frac{m\pi y}{a}\right) \\ &\equiv -\frac{2b^2}{\pi^2 a D}\sum_{m=1}^{\infty}\frac{E_m}{m^3}\sum_{n=1}^{\infty}\frac{(-1)^n n \sin\dfrac{n\pi y}{b}}{\left(\dfrac{b^2}{a^2}+\dfrac{n^2}{m^2}\right)^2}, \\ \text{and similarly,} \\ \left[\frac{\partial w}{\partial x}\right]_{x=a} &= -\frac{2b^2}{\pi^2 a D}\sum_{m=1}^{\infty}\frac{(-1)^m E_m}{m^3}\sum_{n=1}^{\infty}\frac{(-1)^n n \sin\dfrac{n\pi y}{b}}{\left(\dfrac{b^2}{a^2}+\dfrac{n^2}{m^2}\right)^2}. \end{aligned}\right\} \tag{2.70}$$

Similar formulas may now be written down giving the edge slopes due to an arbitrary distribution of moments applied to each of the other edges. The coefficients E_m, and so on, are then determined from the condition of zero edge slope. First, however, the edge slopes of the loaded, simply supported plate are required. When the loading is perfectly general these may be found from the analysis of Section 2.1 in which the deflexion was expressed in the form

$$w = \frac{1}{D\pi^4}\sum_{m=1}^{\infty}\sum_{n=1}^{\infty}\frac{q_{mn}}{\left(\dfrac{m^2}{a^2}+\dfrac{n^2}{b^2}\right)^2}\sin\frac{m\pi x}{a}\sin\frac{n\pi y}{b}, \tag{2.71}$$

where q_{mn} is defined by (2.1).

Differentiating (2.71) gives

$$\left.\begin{aligned}\left[\frac{\partial w}{\partial y}\right]_{y=0} &= \frac{1}{Db\pi^3}\sum_{m=1}^{\infty}\sum_{n=1}^{\infty}\frac{nq_{mn}}{\left(\frac{m^2}{a^2}+\frac{n^2}{b^2}\right)^2}\sin\frac{m\pi x}{a},\\ \left[\frac{\partial w}{\partial y}\right]_{y=b} &= \frac{1}{Db\pi^3}\sum_{m=1}^{\infty}\sum_{n=1}^{\infty}\frac{(-1)^n nq_{mn}}{\left(\frac{m^2}{a^2}+\frac{n^2}{b^2}\right)^2}\sin\frac{m\pi x}{a},\end{aligned}\right\} \qquad (2.72)$$

and there are analogous expressions for $[\partial w/\partial x]_{x=0}$, $[\partial w/\partial x]_{x=a}$.

References

Cox, H. L. *The buckling of plates and shells.* Pergamon, 1962.

Hopkins, H. G. The solution of small displacement, stability or vibration problems concerning a flat rectangular panel when the edges are either clamped or simply supported. *Aero. Res. Council R. and M.* No. 2234. H.M.S.O. (June 1945).

Levy, M. *Compt. rend.*, **129**, pp. 535–9 (1899).

Morley, L. S. D. Simple series solution for the bending of a clamped rectangular plate under uniform normal load. *Quart. J. Mech. Appl. Math.*, **16**, pp. 109–14 (1963).

——. Bending of clamped rectilinear plates. *Quart. J. Mech. Appl. Math.*, **16**, pp. 293–317 (1964).

Nadai, A. *Elastische Platten.* Berlin, 1925.

Thompson, J. M. T. *General theory of elastic stability.* Wiley-Interscience, 1973.

Timoshenko, S. *Theory of elastic stability.* 1st ed. McGraw-Hill, 1936.

——. *Proceedings of the Fifth International Congress in Applied Mechanics,* Cambridge, Mass., 1938.

Timoshenko, S. P., and Woinowsky-Krieger, S. *Theory of plates and shells.* 2nd ed. Chaps. 5, 6, 7. McGraw-Hill, 1959.

3

Plates of various shapes

In this chapter we consider plates of constant thickness whose boundaries are circular or sector-shaped, elliptical, triangular or parallelogram-shaped. Attention is largely confined to the case of isotropy.

3.1 Circular plates

In discussing circular, annular or sector-shaped plates it is advantageous to use polar coordinates as shown in Fig. 3.1. The governing differential equations for the deflexion w and the middle-surface force function Φ are most conveniently derived from those forms, for example (1.64) and (1.65), that are expressed in terms of the invariant operators ∇^2 and $\diamondsuit^4$. We make use of the known relation

$$\nabla^2 \equiv \frac{\partial^2}{\partial r^2} + \frac{1}{r}\frac{\partial}{\partial r} + \frac{1}{r^2}\frac{\partial^2}{\partial \theta^2} \tag{3.1}$$

which enables us, via the definition of $\diamondsuit^4$ after (1.28), to write

$$\diamondsuit^4(\Phi, w) \equiv \frac{\partial^2 \Phi}{\partial r^2}\left(\frac{1}{r}\frac{\partial w}{\partial r} + \frac{1}{r^2}\frac{\partial^2 w}{\partial \theta^2}\right) - 2\frac{\partial}{\partial r}\left(\frac{1}{r}\frac{\partial \Phi}{\partial \theta}\right)\frac{\partial}{\partial r}\left(\frac{1}{r}\frac{\partial w}{\partial \theta}\right) + \left(\frac{1}{r}\frac{\partial \Phi}{\partial r} + \frac{1}{r^2}\frac{\partial^2 \Phi}{\partial \theta^2}\right)\frac{\partial^2 w}{\partial r^2}. \tag{3.2}$$

From Section 1.1 the moments per unit length, M_r, M_θ, $M_{r\theta}$ acting on an element are related to the curvatures by the equations

$$\left.\begin{aligned} M_r &= -D\left(\frac{\partial^2 w}{\partial n^2} + \nu\frac{\partial^2 w}{\partial t^2}\right),\\ M_\theta &= -D\left(\frac{\partial^2 w}{\partial t^2} + \nu\frac{\partial^2 w}{\partial n^2}\right),\\ M_{r\theta} &= -D(1-\nu)\frac{\partial^2 w}{\partial n \partial t}, \end{aligned}\right\} \tag{3.3}$$

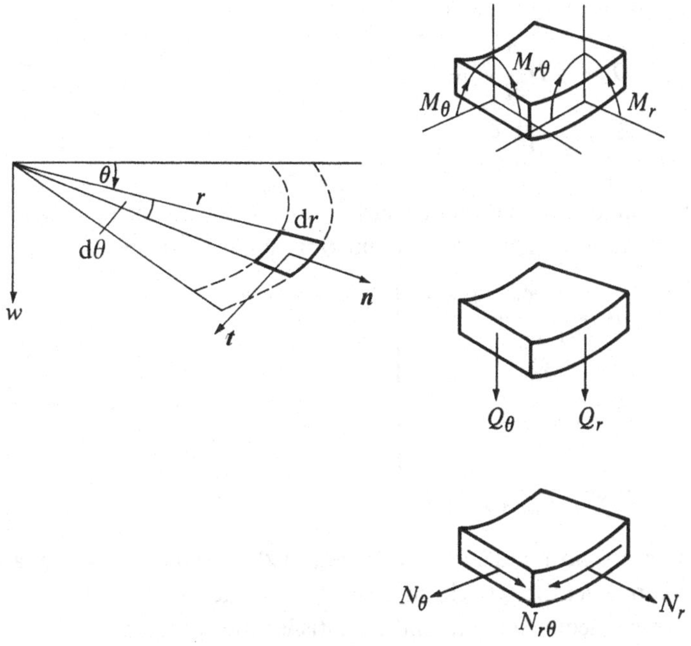

Fig. 3.1

where $\boldsymbol{n}$ and $\boldsymbol{t}$ are measured along the normal and the tangent to the sides of the element. Equation (3.3) may be written in polar coordinates using the known relations

$$\left.\begin{aligned} \frac{\partial^2}{\partial \boldsymbol{n}^2} &\equiv \frac{\partial^2}{\partial r^2}, \\ \frac{\partial^2}{\partial \boldsymbol{t}^2} &\equiv \frac{1}{r}\frac{\partial}{\partial r} + \frac{1}{r^2}\frac{\partial^2}{\partial \theta^2}, \\ \frac{\partial^2}{\partial \boldsymbol{n}\,\partial \boldsymbol{t}} &\equiv \frac{\partial}{\partial r}\left(\frac{1}{r}\frac{\partial}{\partial \theta}\right). \end{aligned}\right\} \tag{3.4}$$

Thus,

$$\left.\begin{aligned} M_r &= -D\left\{\frac{\partial^2 w}{\partial r^2} + \nu\left(\frac{1}{r}\frac{\partial w}{\partial r} + \frac{1}{r^2}\frac{\partial^2 w}{\partial \theta^2}\right)\right\}, \\ M_\theta &= -D\left(\frac{1}{r}\frac{\partial w}{\partial r} + \frac{1}{r^2}\frac{\partial^2 w}{\partial \theta^2} + \nu\frac{\partial^2 w}{\partial r^2}\right), \\ M_{r\theta} &= -D(1-\nu)\frac{\partial}{\partial r}\left(\frac{1}{r}\frac{\partial w}{\partial \theta}\right). \end{aligned}\right\} \tag{3.5}$$

Similarly, we find from (1.31)

$$\left.\begin{aligned} Q_r &= -D\frac{\partial}{\partial r}\nabla^2 w, \\ Q_\theta &= -D\frac{1}{r}\frac{\partial}{\partial \theta}\nabla^2 w. \end{aligned}\right\} \tag{3.6}$$

In the same way, the middle-surface forces per unit length, N_r, N_θ, $N_{r\theta}$, may be derived from the force function Φ by the relations

$$\left.\begin{aligned} N_r &= \frac{1}{r}\frac{\partial \Phi}{\partial r} + \frac{1}{r^2}\frac{\partial^2 \Phi}{\partial \theta^2}, \\ N_\theta &= \frac{\partial^2 \Phi}{\partial r^2}, \\ N_{r\theta} &= -\frac{\partial}{\partial r}\left(\frac{1}{r}\frac{\partial \Phi}{\partial \theta}\right). \end{aligned}\right\} \tag{3.7}$$

If there are no middle-surface forces, (3.2) becomes $D\nabla^4 w = q$, and it is now convenient to search for a solution in the form $w = w_1 + w_2$, where w_1 is a particular integral and w_2 satisfies the equation

$$\nabla^4 w_2 = 0. \tag{3.8}$$

We can find the general solution of (3.8) by taking

$$w_2 = R_0 + \sum_{m=1}^{\infty} R_m \cos\frac{m\pi\theta}{\alpha} + \sum_{m=1}^{\infty} R'_m \sin\frac{m\pi\theta}{\alpha}, \tag{3.9}$$

where α is a constant and the R's are functions of r satisfying the equation

$$\left(\frac{\mathrm{d}^2}{\mathrm{d}r^2} + \frac{1}{r}\frac{\mathrm{d}}{\mathrm{d}r} - \frac{m^2\pi^2}{\alpha^2 r^2}\right)\left(\frac{\mathrm{d}^2 R}{\mathrm{d}r^2} + \frac{1}{r}\frac{\mathrm{d}R}{\mathrm{d}r} - \frac{m^2\pi^2}{\alpha^2 r^2}R\right) = 0. \tag{3.10}$$

The general solution of (3.10) is given by

$$R_m = A_m \varrho^{m\pi/\alpha} + B_m \varrho^{-m\pi/\alpha} + C_m \varrho^{2+m\pi/\alpha} + D_m \varrho^{2-m\pi/\alpha}, \tag{3.11}$$

where $\varrho = r/r_1$ and r_1 is a convenient arbitrary constant. There is a similar expression for R'_m. When m is zero, (3.11) assumes the form

$$R_0 = A_0 + B_0 \ln\varrho + C_0\varrho^2 + D_0\varrho^2\ln\varrho \tag{3.12}$$

and if $m\pi/\alpha = 1$, we find

$$(R)_{m\pi/\alpha=1} = A_1\varrho + B_1\varrho^{-1} + C_1\varrho^3 + D_1\varrho\ln\varrho. \tag{3.13}$$

The angle α has been introduced to facilitate the analysis of plates in the form of a sector subtending an angle α. In the analysis of circular or semi-circular plates we take $\alpha = \pi$.

3.1.1 Plates with rotational symmetry

Such cases were first discussed by Poisson (1829), and a variety of particular solutions are given by Timoshenko and Woinowsky-Krieger (1959). If there is rotational symmetry, the deflexion is independent of θ and the governing differential equation may be cast in the form

$$\frac{1}{\varrho}\frac{\mathrm{d}}{\mathrm{d}\varrho}\left[\varrho\frac{\mathrm{d}}{\mathrm{d}\varrho}\left\{\frac{1}{\varrho}\frac{\mathrm{d}}{\mathrm{d}\varrho}\left(\varrho\frac{\mathrm{d}w}{\mathrm{d}\varrho}\right)\right\}\right]=\frac{r_1^4}{D}q. \tag{3.14}$$

This form lends itself to repeated integration for determining w_1. Thus,

$$w_1=\frac{r_1^4}{D}\int_0^\varrho\frac{1}{\varrho}\left[\int_0^\varrho\varrho\left\{\int_0^\varrho\frac{1}{\varrho}\left(\int_0^\varrho q\varrho\,\mathrm{d}\varrho\right)\mathrm{d}\varrho\right\}\mathrm{d}\varrho\right]\mathrm{d}\varrho. \tag{3.15}$$

For example, if q is constant we obtain

$$w_1=\frac{qr_1^4}{64D}\varrho^4. \tag{3.16}$$

The constants A_0, B_0, C_0, D_0 of (3.12) are now to be determined from the boundary conditions. The radial moments, shears and slopes due to w_1 may be determined in the general case from (3.5), (3.6) and (3.15). Those due to w_2 are determined from (3.5), (3.6) and (3.12):

$$\left.\begin{aligned}(M_r)_2&=\frac{D}{r_1^2}\left(B_0(1-\nu)\frac{1}{\varrho^2}-2C_0(1+\nu)-D_0\{3+\nu+2(1+\nu)\ln\varrho\}\right)\\(Q_r)_2&=\frac{4D\,D_0}{r_1^3\;\varrho},\\\frac{\mathrm{d}w_2}{\mathrm{d}r}&=\frac{1}{r_1}\left(\frac{B_0}{\varrho}+2C_0\varrho+D_0\varrho(1+2\ln\varrho)\right).\end{aligned}\right\} \tag{3.17}$$

Notice that the constant A_0 does not appear in (3.17) because it represents a rigid body movement. Further, if the plate is a complete circle, rather than an annulus, the radial slope at the centre is zero, so that $B_0=0$. The term $D_0\varrho^2\ln\varrho$ in w_2 is the only term that gives rise to a shear $(Q_r)_2$, and by integrating this shear around a circumference we can express the constant D_0 in terms of a total vertical force, P_2, carried by the plate *at every radius*:

$$\begin{aligned}P_2&=2\pi r(Q_r)_2\\&=-\frac{8\pi D}{r_1^2}D_0.\end{aligned} \tag{3.18}$$

If the plate is a complete circle the constant D_0 is, therefore, zero unless there is a concentrated load P_2 acting at the centre.

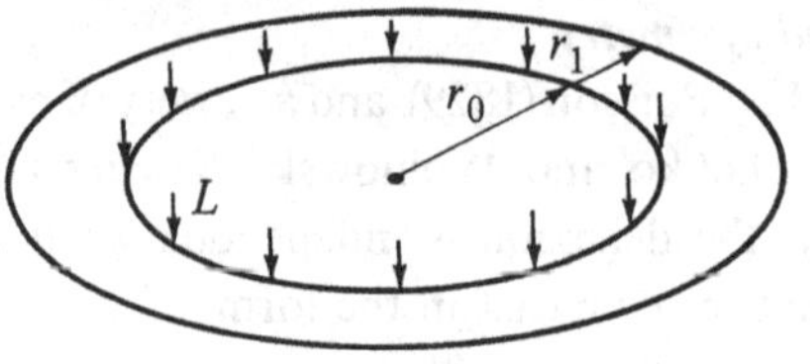

Fig. 3.2

Simply supported plate under uniform load. As a first example, the simply supported plate under uniform load may be considered. The deflexion is given by $w = w_1 + w_2$ where w_1 is given by (3.16) and w_2 by (3.12) with B_0 and D_0 zero. The radial moment M_r is thus given by

$$M_r = -\frac{(3+\nu)qr_1^2}{16}\varrho^2 - \frac{2(1+\nu)C_0 D}{r_1^2}. \tag{3.19}$$

The radius r_1 has still to be chosen and the obvious choice is to equate it to the radius of the plate, so making $\varrho = 1$ at the periphery. The constant C_0 is now determined from the condition of zero M_r at $\varrho = 1$ whence

$$C_0 = -\frac{(3+\nu)qr_1^4}{32(1+\nu)D}. \tag{3.20}$$

The deflexion at the edge may be made zero by taking

$$A_0 = \frac{(5+\nu)qr_1^4}{64(1+\nu)D}. \tag{3.21}$$

Circular plate with ring loading. As a second example, we consider a plate of radius r_1 carrying a total load $2\pi r_0 L$ distributed as a line load of intensity L at a radius r_0, as shown in Fig. 3.2.

Despite the discontinuous character of the applied loading, the deflexion w_1 may still be determined directly from (3.15). However, it is necessary to discriminate between the ranges $0 < \varrho < \varrho_0$ and $\varrho_0 < \varrho < 1$. This will be done by using the symbols $]_0^{\varrho_0}$ and $]_{\varrho_0}^1$. Referring to (3.15), we may then write

$$\int_0^{\varrho} q\varrho \, d\varrho = 0\Big]_0^{\varrho_0} + \frac{Lr_0}{r_1^2}\Big]_{\varrho_0}^1$$

so that

$$\int_0^{\varrho} \frac{1}{\varrho}\left(\int_0^{\varrho} q\varrho \, d\varrho\right) d\varrho = 0\Big]_0^{\varrho_0} + \frac{Lr_0}{r_1^2}\ln\frac{\varrho}{\varrho_0}\Big]_{\varrho_0}^1$$

whence

$$\int_0^\varrho \varrho\left\{\int_0^\varrho \frac{1}{\varrho}\left(\int_0^\varrho q\varrho\,\mathrm{d}\varrho\right)\mathrm{d}\varrho\right\}\mathrm{d}\varrho$$

$$= 0\Big]_0^{\varrho_0} + \frac{Lr_0}{4r_1^2}\left(2\varrho^2\ln\frac{\varrho}{\varrho_0} - \varrho^2 + \varrho_0^2\right)\Big]_{\varrho_0}^1$$

and finally

$$w_1 = \frac{r_1^4}{D}\int_0^\varrho \frac{1}{\varrho}\left[\int_0^\varrho \varrho\left\{\int_0^\varrho \frac{1}{\varrho}\left(\int_0^\varrho q\varrho\,\mathrm{d}\varrho\right)\mathrm{d}\varrho\right\}\mathrm{d}\varrho\right]\mathrm{d}\varrho$$

$$= 0\Big]_0^{\varrho_0} + \frac{Lr_0r_1^2}{4D}\{(\varrho_0^2 + \varrho^2)\ln\frac{\varrho}{\varrho_0} + \varrho_0^2 - \varrho^2\}\Big]_{\varrho_0}^1. \tag{3.22}$$

The complete solution may now be obtained in the usual way by superimposing a deflexion w_2, chosen to satisfy the boundary conditions.

Effect of middle-surface forces. If there is all-around tension or compression such that $N_r = N_\theta = N$, the governing differential equation is obtained from (3.2) by taking

$$\Phi = \tfrac{1}{2}Nr^2 \tag{3.23}$$

which yields

$$\left(\frac{\partial^2}{\partial r^2} + \frac{1}{r}\frac{\partial}{\partial r}\right)\left(\frac{\partial^2 w}{\partial r^2} + \frac{1}{r}\frac{\partial w}{\partial r} - \frac{N}{D}w\right) = \frac{q}{D}. \tag{3.24}$$

The complementary solution of (3.24) is expressed in terms of Bessel functions, depending on the sign of N. Thus, if N is positive (i.e. tensile)

$$\left.\begin{aligned} &w_2 = A_0 + B_0\ln\varrho + C_0 I_0(\beta\varrho) + D_0 K_0(\beta\varrho), \\ &\text{where} \\ &\beta = r_1\sqrt{\frac{N}{D}}, \end{aligned}\right\} \tag{3.25}$$

and I_0, K_0 are Bessel functions of zero order with purely imaginary argument. If N is negative (i.e. compressive),

$$\left.\begin{aligned} &w_2 = A_0 + B_0\ln\varrho + C_0 J_0(\beta'\varrho) + D_0 Y_0(\beta'\varrho), \\ &\text{where} \\ &\beta' = r_1\sqrt{\frac{-N}{D}} \end{aligned}\right\} \tag{3.26}$$

and J_0, Y_0 are Bessel functions of zero order. For further information on Bessel functions see, for example, Whittaker and Watson (1940).

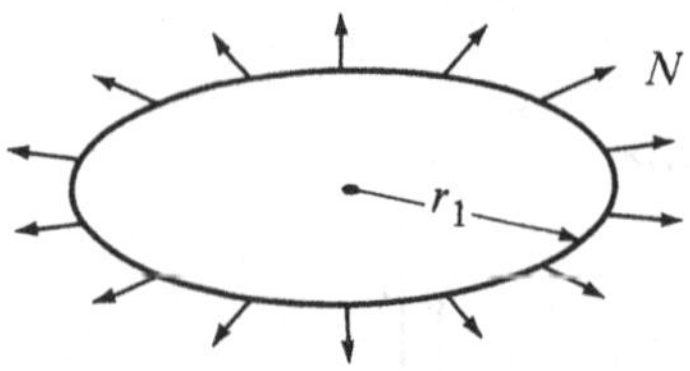

Fig. 3.3

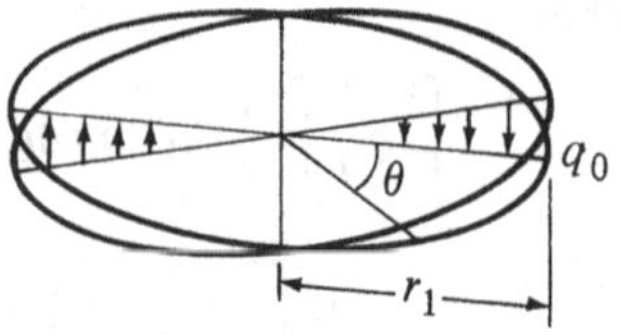

Fig. 3.4

If the plate is a complete circle, rather than an annulus, the central deflexion is finite and it is necessary for B_0 and D_0 above to be zero.

As an example, the uniformly loaded plate under tension is considered (Fig. 3.3). A particular integral of (3.24) is then given by $w_1 = -qr^2/4N$, so that we may take

$$w = \frac{qr_1^2}{4N}((1-\varrho^2) + C_0\{I_0(\beta\varrho) - I_0(\beta)\}) \tag{3.27}$$

which vanishes at $r = r_1$. The constant C_0 is to be determined from the further boundary condition at r_1. Thus if the plate is clamped

$$C_0 = \frac{2}{\beta I_1(\beta)} \tag{3.28}$$

and if it is simply supported

$$C_0 = \frac{2(1+\nu)}{\beta\{\beta I_0(\beta) - (1-\nu)I_1(\beta)\}}. \tag{3.29}$$

3.1.2 *Circular and annular plates under linearly varying load*

These problems were first considered by Flügge (1929), and numerous further examples are given by Timoshenko and Woinowsky-Krieger (1959).

The load distribution shown in Fig. 3.4 is given by

$$q = q_0\varrho\cos\theta \tag{3.30}$$

for which a particular integral of (3.2) is

$$w_1 = \frac{q_0 r_1^4}{192D}\varrho^5\cos\theta. \tag{3.31}$$

All problems of similarly loaded circular and annular plates whose boundary conditions are independent of θ may be solved by combining (3.31) with the complementary solution (3.13). Thus we may write

$$w = \frac{q_0 r_1^4\cos\theta}{192D}\left\{\varrho^5 + A_1\varrho + \frac{B_1}{\varrho} + C_1\varrho^3 + D_1\varrho\ln\varrho\right\} \tag{3.32}$$

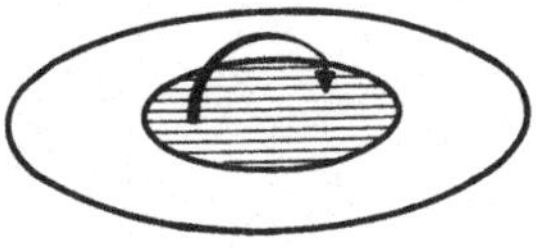

Fig. 3.5

and the constants A_1, B_1, C_1, D_1 are to be determined from the boundary conditions in the usual manner. For a circular plate the deflexion and slope are finite at the origin, so that $B_1 = D_1 = 0$. Note that the boundary condition appropriate to a free edge as given by (1.58) becomes, in polar coordinates,

$$Q_r + \frac{1}{r}\frac{\partial M_{r\theta}}{\partial \theta} = 0. \tag{3.33}$$

Problems in which the loading takes the form of a moment applied to a central rigid disk (Fig. 3.5) may also be solved in a similar way by taking

$$w = \{A_1\varrho + B_1/\varrho + C_1\varrho^3 + D_1\varrho\ln\varrho\}\cos\theta. \tag{3.34}$$

Effect of middle-surface forces. When there is uniform all-around tension (or compression), so that $\Phi = \frac{1}{2}Nr^2$, equation (3.2) becomes

$$\left.\begin{aligned}\nabla^2\left(\nabla^2 w - \frac{N}{D}w\right) &= \frac{q}{D}\\ &= \frac{q_0}{D}\varrho\cos\theta\end{aligned}\right\} \tag{3.35}$$

for a plate under a linearly varying load.

The general solution of (3.35), which gives rise to a deflexion proportional to $\cos\theta$, depends on the sign of N. If N is positive (i.e. tensile),

$$\left.\begin{aligned} w &= \frac{q_0 r_1^2\cos\theta}{8N}\{-\varrho^3 + A_1\varrho + B_1/\varrho + C_1 I_1(\beta\varrho) + D_1 K_1(\beta\varrho)\},\\ &\text{where}\\ \beta &= r_1\sqrt{\frac{N}{D}},\end{aligned}\right\} \tag{3.36}$$

and I_1, K_1 are Bessel functions of order unity with purely imaginary argument. Similarly, if N is negative (i.e. compressive),

$$w = \frac{q_0 r_1^2\cos\theta}{8N}\{-\varrho^3 + A_1\varrho + B_1/\varrho + C_1 J_1(\beta'\varrho) + D_1 Y_1(\beta'\varrho)\},$$

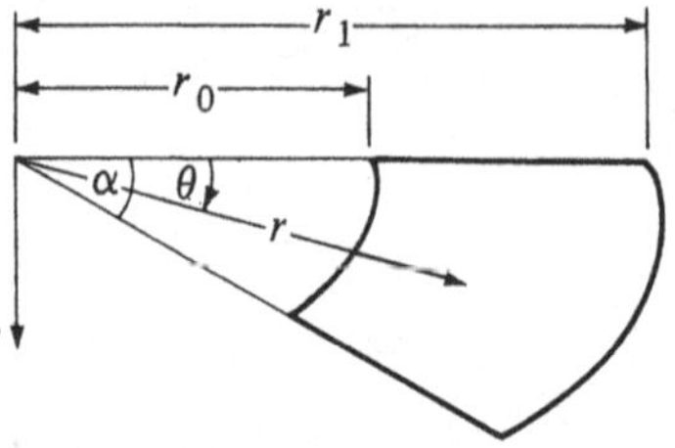

Fig. 3.6

where

$$\beta' = r_1\sqrt{\frac{-N}{D}} \qquad (3.37)$$

and J_1, Y_1 are Bessel functions of order unity.

If the plate is a complete circle the central deflexion is finite and it is necessary for B_1 and D_1 to be zero.

3.2 Uniformly loaded sector plate

We consider a uniformly loaded plate in the form of a sector subtending an angle α and bounded by the lines $r = r_0, r_1$. The plate is assumed to be simply supported along the straight edges (Fig. 3.6).

The uniformly distributed load q_0 may be expressed in the form

$$q = \frac{4q_0}{\pi}\sum_{m=1,3,5}^{\infty}\frac{1}{m}\sin\frac{m\pi\theta}{\alpha}, \qquad (3.38)$$

and a particular integral for the deflexion w_1 which satisfies the conditions of simple support along the edges $\theta = 0, \alpha$ is given by

$$w_1 = \frac{4q_0 r_1^4}{\pi}\varrho^4 \sum_{m=1,3,5}^{\infty}\frac{1}{m(16 - m^2\pi^2/\alpha^2)(4 - m^2\pi^2/\alpha^2)}\sin\frac{m\pi\theta}{\alpha}. \qquad (3.39)$$

If either of the factors $(16 - m^2\pi^2/\alpha^2)$ or $(4 - m^2\pi^2/\alpha^2)$ vanish for a particular value of m (for example, if $\alpha = \frac{1}{2}\pi$ and $m = 1$), it is necessary to search for an alternative term proportional to $\varrho^4 \ln \varrho$. Such a term is given by

$$\frac{4q_0 r_1^4}{\pi}\varrho^4 \ln \varrho\left(\frac{1}{12m(8 - m^2\pi^2/\alpha^2)}\right)\sin\frac{m\pi\theta}{\alpha}. \qquad (3.39a)$$

The complete solution is obtained by combining (3.39) with the sine-terms in (3.9). For example, if $\alpha = \frac{1}{4}\pi$, we may write

$$w = -\frac{q_0 r_1^4}{24\pi D}(\varrho^4 \ln \varrho + A_1\varrho^4 + B_1\varrho^{-4} + C_1\varrho^6 + D_1\varrho^{-2})\sin 4\theta$$

$$+\frac{q_0 r_1^4}{16\pi D}\sum_{m=3,5,7}^{\infty}\frac{\varrho^4+A_m\varrho^{4m}+B_m\varrho^{-4m}+C_m\varrho^{2+4m}+D_m\varrho^{2-4m}}{m(m^2-1)(4m^2-1)}$$
$$\times \sin 4m\theta, \tag{3.40}$$

where the coefficients A_m, B_m, C_m, D_m are chosen to satisfy the boundary conditions along $\varrho=\varrho_0$, 1.

A similar analysis is possible whenever the applied load can be expressed in the form

$$q=r^{\lambda}\sum_{m=1}^{\infty} q_m \sin\frac{m\pi\theta}{\alpha}. \tag{3.41}$$

3.3 Sector and wedge-shaped plates with general boundary conditions

The previous analysis is applicable only to plates simply supported along the straight edges. When these boundary conditions are other than simply supported, it is necessary to derive an alternative representation of the 'general', solution of Section 3.1. This is accomplished by noting that the solution of Section 3.1 is not restricted to integral values for m. A valid solution is also obtained by letting m assume complex values. In particular, if

$$\frac{m\pi}{\alpha}=1\pm iu, \tag{3.42}$$

we derive the following expression for w_2:

$$\left.\begin{aligned}
w_2&=\varrho\left[\Theta_0+\sum_{u=1}^{\infty}\Theta_u\cos(u\ln\varrho)+\sum_{u=1}^{\infty}\Theta'_u\sin(u\ln\varrho)\right],\\
&\text{where}\\
\Theta_0&=a_0\cos\theta+b_0\sin\theta+c_0\theta\cos\theta+d_0\theta\sin\theta,\\
\Theta_u&=a_u\cosh u\theta\cos\theta+b_u\cosh u\theta\sin\theta\\
&\quad+c_u\sinh u\theta\cos\theta+d_u\sinh u\theta\sin\theta,\\
\Theta'_u&=a'_u\cosh u\theta\cos\theta+b'_u\cosh u\theta\sin\theta\\
&\quad+c'_u\sinh u\theta\cos\theta+d'_u\sinh u\theta\sin\theta.
\end{aligned}\right\} \tag{3.43}$$

If the deflexion is zero at $\varrho=\varrho_0$, 1 it is preferable to take

$$u=\frac{v\pi}{\ln\varrho_0}, \tag{3.44}$$

where v assumes positive integral values. The deflexion w_2 may then be written

$$w_2=\varrho\sum_{v=1}^{\infty}\Theta_v\sin\left(\frac{v\pi\ln(\varrho/\varrho_0)}{\ln\varrho_0}\right). \tag{3.45}$$

Further, if w_2 is symmetrical about the line $\theta = 0$ and zero along $\theta = \pm\frac{1}{2}\alpha$, say, the function Θ_v is given by

$$\Theta_v = A_v \left(\frac{\cosh\left(\dfrac{\pi v \theta}{\ln \varrho_0}\right) \cos\theta}{\cosh\left(\dfrac{\pi v \alpha}{2 \ln \varrho_0}\right) \cos\frac{1}{2}\alpha} - \frac{\sinh\left(\dfrac{\pi v \theta}{\ln \varrho_0}\right) \sin\theta}{\sinh\left(\dfrac{\pi v \alpha}{2 \ln \varrho_0}\right) \sin\frac{1}{2}\alpha} \right). \tag{3.46}$$

By combining solutions of this form with those obtained in the previous sections, it is possible to satisfy general boundary conditions. The method is analogous to that outlined in Section 2.3 for the clamped rectangular plate. The uniformly loaded clamped sector was first solved by Carrier (1944), while more exact solutions, based on variational methods, were derived by Morley (1963) (see Section 6.6.1).

3.4 Clamped elliptical plate

Simple solutions exist for the clamped elliptical plate under a uniform distribution of load and under a linearly varying load. It was shown by Bryan that if

$$q = q_0 + q_1 x/a, \tag{3.47}$$

the deflexion is given by

$$\left.\begin{aligned} w &= \frac{a^4}{8D}\left(1 - \frac{x^2}{a^2} - \frac{y^2}{b^2}\right)^2 \left(\frac{q_0}{3 + 2\zeta^2 + 3\zeta^4} + \frac{q_1 x/a}{3(5 + 2\zeta^2 + \zeta^4)}\right), \\ &\text{where} \\ \zeta &= a/b. \end{aligned}\right\} \tag{3.48}$$

3.4.1 *Effect of anisotropy*

In what follows we concentrate on the class of anisotropy discussed in Section 1.8.3. Then, for the case of uniform load q_0, we find that (1.101) is satisfied by a deflexion that differs only in magnitude from that for the isotropic plate:

$$\left.\begin{aligned} w &= w_0\left(1 - \frac{x^2}{a^2} - \frac{y^2}{b^2}\right)^2, \\ &\text{where} \\ w_0 &= \frac{a^4 q_0}{8\{3D_{11} + 2\zeta^2(D_{12} + 2D_{66}) + 3\zeta^4 D_{22}\}}. \end{aligned}\right\} \tag{3.49}$$

A linearly varying load, however, results in a deflexion that differs in both form and magnitude from that of (3.48). Thus if

$$q = q_1 x/a,$$

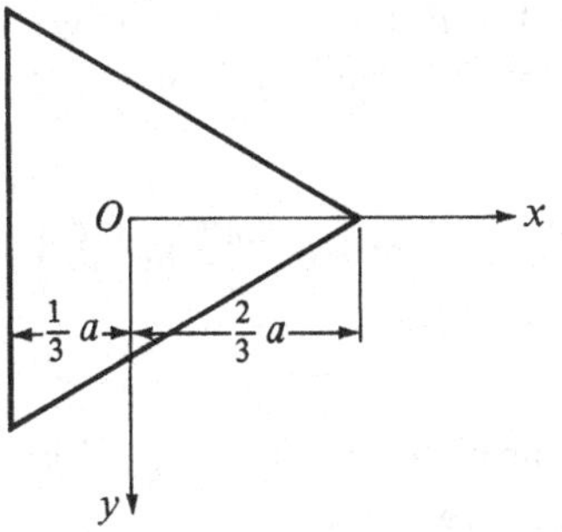

Fig. 3.7

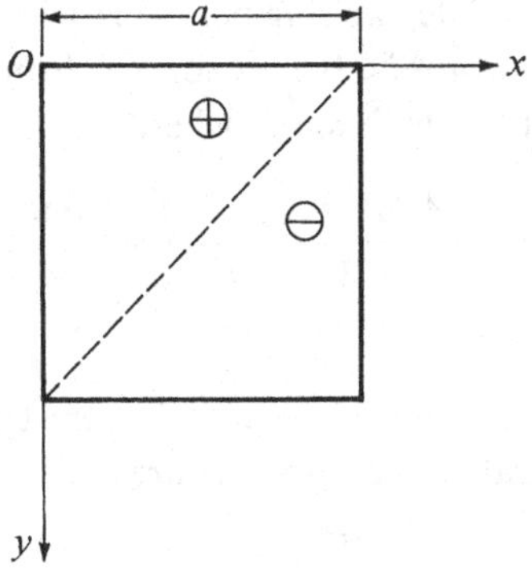

Fig. 3.8

it may be shown that

$$\left.\begin{aligned} w &= \frac{a^3 q_1}{24}(\lambda x + \mu y)\left(1 - \frac{x^2}{a^2} - \frac{y^2}{b^2}\right)^2, \\ &\text{where} \\ \lambda &= \left(\alpha - \frac{16\zeta^2\gamma^2}{\beta}\right)^{-1}, \\ \mu &= -4\lambda\zeta^2\gamma/\beta, \\ &\text{and} \\ \alpha &= 5D_{11} + 2\zeta^2(D_{12} + 2D_{66}) + \zeta^4 D_{22}, \\ \beta &= D_{11} + 2\zeta^2(D_{12} + 2D_{66}) + 5\zeta^4 D_{22}, \\ \gamma &= D_{16} + \zeta^2 D_{26}. \end{aligned}\right\} \tag{3.50}$$

3.5 Simply supported equilateral triangular plate

The deflexion of a simply supported equilateral triangular plate under a uniform loading q_0 has (Fig. 3.7) been shown by Woinowsky-Krieger (1933) to be given by

$$w = \frac{q_0}{64aD}[x^3 - 3y^2x - a(x^2 + y^2) + \tfrac{4}{27}a^3](\tfrac{4}{9}a^2 - x^2 - y^2). \tag{3.51}$$

3.6 Simply supported isosceles right-angled triangular plate

As shown by Nadai (1925), the deflexion of a simply supported plate in the form of an isosceles right-angled triangle may be obtained by using the method of Section 2.1. The analysis is applied to a square plate as shown in Fig. 3.8, the loading on the triangular plate being augmented by a loading which is a 'negative mirror image' about the common diagonal. Thus if there is a concentrated load P at the point (ξ, η) in the triangular plate, we augment this with a load $-P$ at the mirror-image point $(a-\eta, a-\xi)$. It now follows from Section 2.1 that the deflexion is given by (2.6), with $a=b$, where

$$w_{mn}=\frac{4Pa^2}{\pi^4 D}\frac{1}{(m^2+n^2)^2}\left(\sin\frac{m\pi\xi}{a}\sin\frac{n\pi\eta}{a}-(-1)^{m+n}\sin\frac{m\pi\eta}{a}\sin\frac{n\pi\xi}{a}\right) \quad \cdots(3.52)$$

If there is a uniform loading q_0 on the triangular plate, we put $P=q_0\delta\xi\delta\eta$ and integrate (3.52) over the area of the triangle to obtain

$$w=\frac{16qa^4}{\pi^4 D}\left[\sum_{m=1,3,5,\ldots}^{\infty}\sum_{n=2,4,6,\ldots}^{\infty}\frac{n\sin\frac{m\pi x}{a}\sin\frac{n\pi y}{a}}{m(n^2-m^2)(m^2+n^2)^2}\right.$$
$$\left.+\sum_{m=2,4,6,\ldots}^{\infty}\sum_{n=1,3,5,\ldots}^{\infty}\frac{m\sin\frac{m\pi x}{a}\sin\frac{n\pi y}{a}}{n(m^2-n^2)(m^2+n^2)^2}\right]. \quad (3.53)$$

This method can also be applied when the triangular plate is under uniform all-round tension or compression.

3.7 Clamped parallelogram plate

Morley (1963) presented a single-series solution for a clamped isotropic parallelogram plate under uniform normal loading q_0. Each term of the series satisfies the biharmonic equation and the boundary conditions on one pair of opposite edges, in a manner similar to that considered in Section 2.2. The remaining boundary conditions are satisfied by invoking a variational principle of minimum energy introduced by Diaz and Greenberg (1948). It would therefore have been equally appropriate to present this solution in Chapter 6. However, the series expansion chosen for the deflexion is similar to a Fourier expansion in that it can represent an arbitrary function and it converges on the exact solution in a manner similar to that discussed in Section 2.2.

An oblique Cartesian coordinate system $\mathbf{x}O\mathbf{y}$ is chosen with the clamped edges of the plate situated at $\mathbf{x}=\pm a$, $\mathbf{y}=\pm b$; a rectangular coordinate system xOy is superimposed with a common origin and with Ox and $O\mathbf{x}$ coincident (see Fig. 3.9).
The distance b' is given by

$$b'=b\sin\alpha \quad (3.54)$$

and, from (1.67), (1.68), the clamped boundary conditions can be expressed in the form

$$w=\partial w/\partial\mathbf{x}=0 \quad \text{at} \quad \mathbf{x}=\pm a \quad (3.55)$$

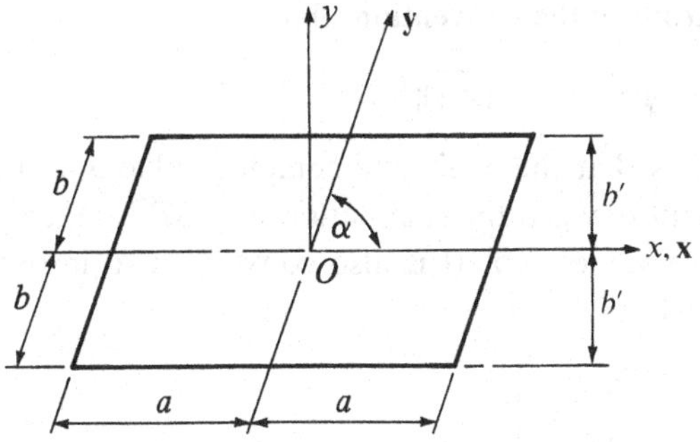

Fig. 3.9

and

$$w = \partial w/\partial \mathrm{y} = 0 \quad \text{at} \quad \mathrm{y} = \pm b. \tag{3.56}$$

We now search for a solution of (1.30) in the form

$$w = w_0 + \sum_r A_r w_r(x, y), \tag{3.57}$$

where w_0 is the particular solution

$$w_0 = \frac{(b'^2 - y^2)^2}{24D} q_0 \tag{3.58}$$

which satisfies the boundary conditions (3.56). The $w_r(x, y)$ are chosen as a sequence of biharmonic functions in which for odd r we put

$$w_r(x, y) = \frac{q_0}{D\lambda_r}(y \sin \lambda_r y - b' \tan \lambda_r b' \cos \lambda_r y) \cosh \lambda_r x \tag{3.59}$$

so that when the constants λ_r are calculated as the (complex) roots of the transcendental equation

$$2\lambda_r b' + \sin 2\lambda_r b' = 0, \tag{3.60}$$

then each function $w_r(x, y)$ satisfies the boundary conditions given in (3.56). Similarly, for even r we put

$$w_r(x, y) = \frac{q_0}{D\lambda_r}(y \cos \lambda_r y - b' \cot \lambda_r b' \sin \lambda_r y) \sin \lambda_r x, \tag{3.61}$$

where the λ_r are now the roots of the transcendental equation

$$2\lambda_r b' - \sin 2\lambda_r b' = 0, \tag{3.62}$$

so that the boundary conditions given in (3.56) are again satisfied. Note that the above sequences are infinite and satisfy the condition of polar symmetry

$$w_r(x, y) = w_r(-x, -y). \tag{3.63}$$

At this stage we introduce the convention that

$$\lambda_{-r} = \bar{\lambda}_r; \quad w_{-r}(x, y) = \overline{w_r(x, y)}; \quad \text{etc.} \tag{3.64}$$

where the bar denotes that the conjugate complex value is to be taken. Thus, w is a real quantity when the summation in (3.57) is taken over the positive and negative values of r. It is also convenient to introduce the complex variable z where

$$\left.\begin{aligned} z &= x + iy = \mathbf{x} + \mathbf{y}e^{i\alpha}, \\ \bar{z} &= x - iy = \mathbf{x} + \mathbf{y}e^{-i\alpha}. \end{aligned}\right\} \tag{3.65}$$

Furthermore, we later require expressions for the Laplacian of w_0 and w_r, namely

$$\nabla^2 w_0 = -\frac{(b'^2 - 3y^2)}{6D} q_0, \tag{3.66}$$

and, when r is odd,

$$\begin{aligned} \nabla^2 w_r &= \frac{2q_0}{D} \cos \lambda_r y \cosh \lambda_r x \\ &= \frac{q_0}{D} (\cosh \lambda_r z + \cosh \lambda_r \bar{z}), \end{aligned} \tag{3.67}$$

while when r is even,

$$\begin{aligned} \nabla^2 w_r &= -\frac{2q_0}{D} \sin \lambda_r y \sinh \lambda_r x \\ &= i\frac{q_0}{D} (\cosh \lambda_r z - \cosh \lambda_r \bar{z}). \end{aligned} \tag{3.68}$$

The variational principle

The physical basis of the variational principle introduced by Diaz and Greenberg (1948) stems from the fact that in a plate with certain 'natural' boundary conditions – for example, simply supported or clamped – no work is done on or by the boundary supports. The work done by the applied loads is thus equal to the strain energy stored in the plate. This strain energy is a minimum when all the boundaries are clamped. Thus from Section 6.1.1 and, in particular, equation (6.62), the remaining boundary condition (3.55) can be satisfied and the coefficients A_r determined by minimizing the double integral

$$\int_{-b}^{b} \int_{-a}^{a} \{\nabla^2 w(\mathbf{x}, \mathbf{y})\}^2 \, d\mathbf{x} \, d\mathbf{y}. \tag{3.69}$$

We now introduce the following notation

$$\left.\begin{aligned} V(w_r, w_s) = V(w_s, w_r) &= \int_{-b}^{b}\int_{-a}^{a} \nabla^2 w_r \nabla^2 w_s \, \mathbf{dx}\, \mathbf{dy}, \\ V(w_r, w_0) &= -\int_{-b}^{b}\int_{-a}^{a} \nabla^2 w_r \nabla^2 w_0 \, \mathbf{dx}\, \mathbf{dy}, \end{aligned}\right\} \tag{3.70}$$

where in virtue of the simplicity of (3.66)–(3.68) the integrals are very easily evaluated. They are

$$V(w_r, w_s) = \frac{q_0^2}{D^2}\{I(\lambda_r z, \lambda_s z) + I(\lambda_r \bar{z}, \lambda_s \bar{z}) + I(\lambda_r z, \lambda_s \bar{z}) + I(\lambda_r \bar{z}, \lambda_s z)\} \tag{3.71}$$

when both r and s are odd;

$$V(w_r, w_s) = -\frac{q_0^2}{D^2}\{I(\lambda_r z, \lambda_s z) + I(\lambda_r \bar{z}, \lambda_s \bar{z}) - I(\lambda_r z, \lambda_s \bar{z}) - I(\lambda_r \bar{z}, \lambda_s z)\} \tag{3.72}$$

when both r and s are even;

$$V(w_r, w_s) = i\frac{q_0^2}{D^2}\{I(\lambda_r z, \lambda_s z) - I(\lambda_r \bar{z}, \lambda_s \bar{z}) - (\lambda_r z, \lambda_s \bar{z}) + I(\lambda_r \bar{z}, \lambda_s \bar{z})\} \tag{3.73}$$

when r is odd and s is even. The quantities $I(\lambda_r z, \lambda_s z)$ and $I(\lambda_r z, \lambda_s \bar{z})$ are given by

$$\begin{aligned} I(\lambda_r z, \lambda_s z) &= \int_{-b}^{b}\int_{-a}^{a} \cosh \lambda_r z \cosh \lambda_s z \, \mathbf{dx}\, \mathbf{dy} \\ &= \left[\frac{\cosh(\lambda_r z + \lambda_s z)}{(\lambda_r + \lambda_s)(\lambda_r e^{i\alpha} + \lambda_s e^{i\alpha})} \right. \\ &\quad \left. + \frac{\cosh(\lambda_r z - \lambda_s z)}{(\lambda_r - \lambda_s)(\lambda_r e^{i\alpha} - \lambda_s e^{i\alpha})}\right]_{z=a-be^{i\alpha}}^{z=a+be^{i\alpha}}, \end{aligned} \tag{3.74}$$

except when $r = s$ and then

$$I(\lambda_r z, \lambda_r z) = 2ab + \left[\frac{\cosh 2\lambda_r z}{4\lambda_r^2 e^{i\alpha}}\right]_{z=a-be^{i\alpha}}^{z=a+be^{i\alpha}}; \tag{3.75}$$

and

$$\begin{aligned} I(\lambda_r z, \lambda_s \bar{z}) &= \int_{-b}^{b}\int_{-a}^{a} \cosh \lambda_r z \cosh \lambda_s \bar{z} \, \mathbf{dx}\, \mathbf{dy} \\ &= \left[\frac{\cosh(\lambda_r z + \lambda_s \bar{z})}{(\lambda_r + \lambda_s)(\lambda_r e^{i\alpha} + \lambda_s e^{-i\alpha})} \right. \end{aligned}$$

$$+\frac{\cosh(\lambda_r z-\lambda_s\bar{z})}{(\lambda_r-\lambda_s)(\lambda_r e^{i\alpha}+\lambda_s e^{-i\alpha})}\Bigg]_{z=a-be^{i\alpha}}^{z=a+be^{i\alpha}}, \tag{3.76}$$

except when $r=s$ and then

$$I(\lambda_r z,\lambda_s\bar{z})=\left[\frac{a\sin 2\lambda_r b'}{\lambda_r\sin\alpha}+\frac{\cosh\lambda_r(z+\bar{z})}{4\lambda_r^2\cos\alpha}\right]_{z=a-be^{i\alpha}}^{z=a+be^{i\alpha}}. \tag{3.77}$$

The remaining relationships can be obtained with the help of the identities

$$\left.\begin{aligned}I(\lambda_r\bar{z},\lambda_s\bar{z})&\equiv\overline{I(\lambda_{-r}z,\lambda_{-s}z)},\\ I(\lambda_r\bar{z},\lambda_s z)&\equiv I(\lambda_s z,\lambda_r\bar{z}).\end{aligned}\right\} \tag{3.78}$$

Finally,

$$V(w_r,w_0)=\frac{q_0^2}{D^2}\sin^2\alpha\{I(\lambda_r z)+I(\lambda_r\bar{z})\} \tag{3.79}$$

when r is odd and

$$V(w_r,w_0)=i\frac{q_0^2}{D^2}\sin^2\alpha\{I(\lambda_r z)-I(\lambda_r\bar{z})\} \tag{3.80}$$

when r is even, where

$$\begin{aligned}I(\lambda_r z)&=\int_{-b}^{b}\int_{-a}^{a}\frac{(b^2-3y^2)}{6}\cosh\lambda_r z\,\mathrm{d}x\,\mathrm{d}y\\ &=\frac{4\sinh\lambda_r a}{\lambda_r^4 e^{3i\alpha}}\Big\{b\lambda_r e^{i\alpha}\cosh(\lambda_r be^{i\alpha})\\ &\quad-\left(1+\frac{b^2}{3}\lambda_r^2e^{2i\alpha}\right)\sinh(\lambda_r be^{i\alpha})\Big\}\end{aligned} \tag{3.81}$$

and

$$\begin{aligned}I(\lambda_r\bar{z})&=\int_{-b}^{b}\int_{-a}^{a}\frac{(b^2-3y^2)}{6}\cosh\lambda_r\bar{z}\,\mathrm{d}x\,\mathrm{d}y\\ &=\frac{4\sinh\lambda_r a}{\lambda_r^4 e^{-3i\alpha}}\Big\{b\lambda_r e^{-i\alpha}\cosh(\lambda_r be^{-i\alpha})\\ &\quad-\left(1+\frac{b^2}{3}\lambda_r^2e^{-2i\alpha}\right)\sinh(\lambda_r be^{-i\alpha})\Big\}.\end{aligned} \tag{3.82}$$

In a practical calculation, the infinite series of (3.57) is terminated after the first few terms, say, when $-n\leqslant r\leqslant n$. The minimization of (3.69) then provides the following n complex simultaneous equations for the determination of the n complex coefficients A_r,

$$\begin{bmatrix} V(w_1,\bar{w}_1) & V(w_1,\bar{w}_2) & \cdots\cdots & V(w_1,\bar{w}_n) \\ V(w_2,\bar{w}_1) & V(w_2,\bar{w}_2) & \cdots\cdots & V(w_2,\bar{w}_n) \\ \cdots & \cdots & \cdots & \cdots \\ V(w_n,\bar{w}_1) & V(w_n,\bar{w}_2) & \cdots\cdots & V(w_n,\bar{w}_n) \end{bmatrix} \begin{bmatrix} \bar{A}_1 \\ \bar{A}_2 \\ \cdots \\ \bar{A}_n \end{bmatrix}$$
$$+ \begin{bmatrix} V(w_1,w_1) & V(w_1,w_2) & \cdots\cdots & V(w_1,w_n) \\ V(w_2,w_1) & V(w_2,w_2) & \cdots\cdots & V(w_2,w_n) \\ \cdots & \cdots & \cdots & \cdots \\ V(w_n,w_1) & V(w_n,w_2) & \cdots\cdots & V(w_n,w_n) \end{bmatrix} \begin{bmatrix} A_1 \\ A_2 \\ \cdots \\ A_n \end{bmatrix}$$
$$= \begin{bmatrix} V(w_1,w) \\ V(w_2,w) \\ \cdots\cdots \\ V(w_n,w) \end{bmatrix}. \tag{3.83}$$

Numerical results

The following table giving the displacement w_{max} at the centre of various uniformly loaded clamped parallelogram plates is extracted from Morley (1963). The results for $\alpha = 75°$ demonstrate the rapid convergence of the series expansion (3.57).

$$w_{max} = \frac{cq_0a^4}{D}10^{-2}$$

α degrees	Number n of terms in series	$a/b = 1$ c	$a/b = 1.25$ c	$a/b = 1.5$ c	$a/b = 2.0$ c
75	2	1.872	1.070	0.613	0.221
75	4	1.803	1.059	0.612	0.222
75	6	1.793	1.057	0.612	0.222
75	8	1.792			
60	4		0.734	0.411	0.145
45	4			0.197	0.0652

3.8 Singular behaviour at corners

The efficacy of the above solution for the clamped parallelogram plate stems, in part, from the absence of any singular behaviour in the bending moments at the corners. However, Williams (1951) showed that such singularities may occur for other homogeneous boundary conditions and, as shown in the solution for the simply supported rhombic plate in Section 3.9, particular attention must then be paid to this feature. Consider, therefore, conditions near a corner where the boundaries meet at an angle α. It may be verified that the complementary function for the deflexion includes terms of the form

$$w = r^{\lambda+1}\{C_1 \sin(\lambda+1)\theta + C_2 \cos(\lambda+1)\theta + C_3 \sin(\lambda-1)\theta + C_4 \cos(\lambda-1)\theta\}, \tag{3.84}$$

where λ, C_1, C_2, C_3 and C_4 are constants to be determined from the boundary conditions along the radial edges at $\theta=0$ and $\theta=\alpha$. For example, if both edges are simply supported, we derive the following four simultaneous equations for the constants C_1 to C_4:

$$\begin{bmatrix} 0 & 1 & 0 & 1 \\ \sin(\lambda+1)\alpha & \cos(\lambda+1)\alpha & \sin(\lambda-1)\alpha & \cos(\lambda-1)\alpha \\ 0 & 0 & 0 & 1 \\ 0 & 0 & \sin(\lambda-1)\alpha & \cos(\lambda-1)\alpha \end{bmatrix} \begin{bmatrix} C_1 \\ C_2 \\ C_3 \\ C_4 \end{bmatrix} = 0. \tag{3.85}$$

Non-trivial solutions of these equations occur when the determinant vanishes, that is, when λ is a root of the transcendental equation

$$\sin(\lambda+1)\alpha \sin(\lambda-1)\alpha = 0. \tag{3.86}$$

It may now be shown that the smallest value of λ, consistent with finite values of $\partial w/\partial r$ as $r \to 0$, is given by

$$\lambda = \frac{\pi}{\alpha} - 1. \tag{3.87}$$

Other homogeneous boundary conditions can be treated in a similar manner and the table below (in which s.s. stands for simply supported) gives the corresponding transcendental equation for λ. Some of these equations yield complex values of λ and the character of the moments at the vertex is then determined by that value of λ having the smallest real part, Re λ min.

Case number	Boundary condition $\theta=0$	$\theta=\alpha$	Transcendental equation for λ
(i)	s.s.	s.s.	$\sin(\lambda+1)\alpha \sin(\lambda-1)\alpha = 0$
(ii)	clamped	free	$(3+\nu)(1-\nu)\sin^2\lambda\alpha = 4-(1-\nu)^2\lambda^2\sin^2\alpha$
(iii)	s.s.	free	$(3+\nu)\sin 2\lambda\alpha = -(1-\nu)\lambda\sin 2\alpha$
(iv)	clamped	s.s.	$\sin 2\lambda\alpha = \lambda \sin 2\alpha$
(v)	free	free	$(3+\nu)^2\sin^2\lambda\alpha = (1-\nu)^2\lambda^2\sin^2\alpha$
(vi)	clamped	clamped	$\sin^2\lambda\alpha = \lambda^2\sin^2\alpha$

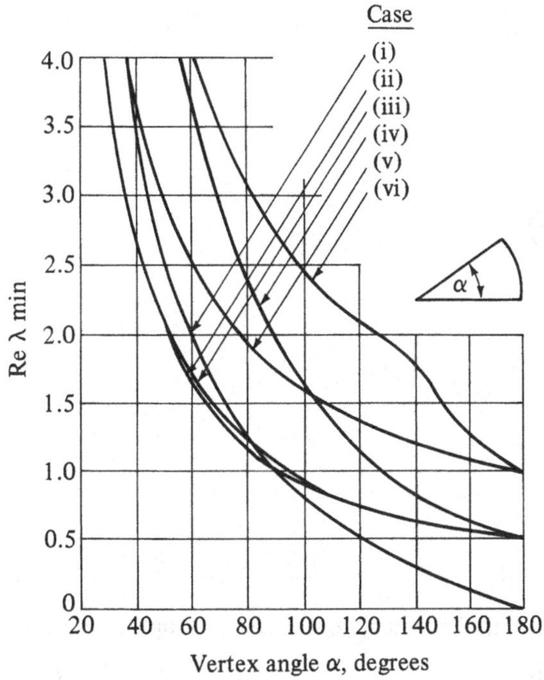

Fig. 3.10 [Extracted from Williams (1951) with the kind permission of the publisher, The American Society of Mechanical Engineers.]

The corresponding variations of $\mathrm{Re}\,\lambda$ min with vertex angle α are shown in Fig. 3.10, assuming $\nu = 0.3$.

If the vertex angle $\alpha \leqslant 90°$ it is seen that $\mathrm{Re}\,\lambda \geqslant 1$ for all the boundary conditions considered and hence there is no singularity in the moments. Singularities may occur, however, for higher values of α; in particular, if both edges are simply supported singularities may occur whenever the vertex angle is obtuse.

3.9 Simply supported rhombic plate

The following solution for the simply supported rhombic plate under a uniformly distributed load q_0 was given by Morley (1962). Because of symmetry in the planform and loading, attention is focused on half the plate; further, because singular behaviour is to be expected at the obtuse vertices, we introduce a polar coordinate system as shown in Fig. 3.11 and we consider the deflexion in the region OAB. Because of the need to express conditions of continuity along the diagonal AB, we also introduce the rectangular Cartesian system as shown.

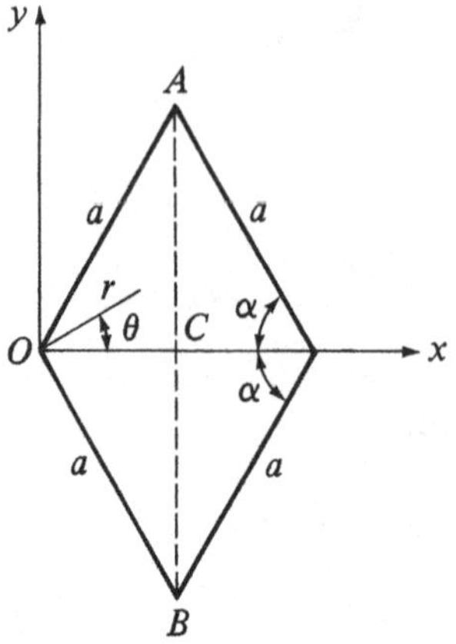

Fig. 3.11

Boundary conditions

Along the sides OA and OB the boundary conditions for simple support are

$$w = \partial^2 w/\partial\theta^2 = 0, \tag{3.88}$$

while along the diagonal AB there is continuity of slope and zero shear resultant, whence

$$\partial w/\partial x = \partial^3 w/\partial x^3 = 0, \tag{3.89}$$

and we note that in polar coordinates

$$\frac{\partial}{\partial x} = \cos\theta \frac{\partial}{\partial r} - \frac{\sin\theta}{r}\frac{\partial}{\partial\theta}.$$

We now search for a solution of the biharmonic equation in the form

$$w = w_0 + w_1, \tag{3.90}$$

where w_1 is a complementary function and w_0 is the particular solution

$$w_0 = \frac{q_0 r^4}{64D}\left(1 - \frac{4}{3}\frac{\cos 2\theta}{\cos 2\alpha} + \frac{1}{3}\frac{\cos 4\theta}{\cos 4\alpha}\right), \tag{3.91}$$

which satisfies the boundary conditions of (3.88) along the edges $\theta = \pm\alpha$. Note that when $\alpha = 3\pi/8$, equation (3.91) is not valid because of the vanishing of a cosine term in the denominator and an alternative expression must then be used (see Morley 1963).

The boundary conditions along the radial edges $\theta = \pm\alpha$ to be satisfied by the function w_1 are therefore

$$w_1 = \partial^2 w_1/\partial\theta^2 = 0, \tag{3.92}$$

while from (3.89)–(3.91) we find that

$$\left[\frac{\partial w_1}{\partial x}\right]_{x=a\cos\alpha} = -\frac{q_0}{16D}\left[\left(1-\frac{1}{\cos 2\alpha}\right)r^3\cos\theta + \frac{1}{3}\left(\frac{1}{\cos 4\alpha}-\frac{1}{\cos 2\alpha}\right)r^3\cos 3\theta\right]_{x=a\cos\alpha} \tag{3.93}$$

and

$$\left[\frac{\partial^3 w_1}{\partial x^3}\right]_{x=a\cos\alpha} = -\frac{q_0 a\cos\alpha}{2D}\left(1-\frac{1}{\cos 2\alpha}\right). \tag{3.94}$$

Following the analysis of Section 3.8, and with a slight change of notation, we search for a solution in the form

$$w_1 = \frac{q_0}{D}\sum_m (a_m + b_m r^2) r^{\lambda_m+1}\cos(\lambda_m+1)\theta, \tag{3.95}$$

where the summation is taken over the positive integral values of m and the λ_m are chosen so that

$$\lambda_m = \frac{(2m-1)\pi}{2\alpha} - 1, \tag{3.96}$$

thus satisfying the boundary conditions (3.92). The coefficients a_m and b_m are to be chosen to satisfy the boundary conditions (3.93), (3.94) and we note that it is the term in a_1 which governs the singular behaviour in the moments as $r \to 0$. When (3.95) is substituted into (3.94), we obtain the following equation which does not contain any of the coefficients a_m,

$$\frac{4q_0}{D}\left[\sum_{m=1}^{\infty}(\lambda_m+2)(\lambda_m+1)b_m r^{\lambda_m}\cos\lambda_m\theta\right]_{x=a\cos\alpha} = -\frac{q_0 a\cos\alpha}{2D}\left(1-\frac{1}{\cos 2\alpha}\right). \tag{3.97}$$

It follows that this equation alone suffices to determine the coefficients b_m. However, (3.97) cannot be solved exactly and we must resort to an approximation whereby the first few coefficients $b_1, b_2, \ldots, b_M$ are determined by the method of least squares, that is,

$$\delta\int_{-a\sin\alpha}^{a\sin\alpha}\left[\frac{\partial^3 w}{\partial x^3}\right]^2_{x=a\cos\alpha} dy = 0. \tag{3.98}$$

This leads to the following system of M linear simultaneous equations

$$\left.\begin{aligned} A_{11}b_1 + A_{12}b_2 + A_{13}b_3 + \cdots + A_{1M}b_M &= A_1 \\ A_{21}b_1 + A_{22}b_2 + A_{23}b_3 + \cdots + A_{2M}b_M &= A_2 \\ A_{31}b_1 + A_{32}b_2 + A_{33}b_3 + \cdots + A_{3M}b_M &= A_3 \\ \cdots\cdots\cdots\cdots\cdots\cdots\cdots \\ A_{M1}b_1 + A_{M2}b_2 + A_{M3}b_3 + \cdots + A_{MM}b_M &= A_M, \end{aligned}\right\} \tag{3.99}$$

where the coefficients A_{mn} are given by

$$\frac{A_{mn}}{16(\lambda_m+2)(\lambda_m+1)(\lambda_n+2)(\lambda_n+1)} = \int_{-a\sin\alpha}^{a\sin\alpha} [r^{\lambda_m+\lambda_n}\cos\lambda_m\theta\cos\lambda_n\theta]_{x=a\cos\alpha}\,dy \tag{3.100}$$

and the A_m by

$$\begin{aligned} \frac{A_m}{4(\lambda_m+2)(\lambda_m+1)} &= \frac{1}{2}\left(\frac{1}{\cos 2\alpha}-1\right)\int_{-a\sin\alpha}^{a\sin\alpha} [r^{\lambda_m+1}\cos\lambda_m\theta\cos\theta]_{x=a\cos\alpha}\,dy \\ &= (-1)^{m+1}(1-\cos 2\alpha)a^{\lambda_m+2}/(\lambda_m+1). \end{aligned} \tag{3.101}$$

It remains now to determine the values of the coefficients, $a_1, a_2, \ldots, a_M$ from the condition obtained by substituting (3.95) into (3.93), that is,

$$\begin{aligned} \frac{q_0}{D}\sum_{m=1}^{M} [(\lambda_m+1)a_m r^{\lambda_m}\cos\lambda_m\theta]_{x=a\cos\alpha} &= \frac{q_0}{16D}\left[\left(1-\frac{1}{\cos 2\alpha}\right)r^3\cos\theta \right. \\ &\quad \left. +\frac{1}{3}\left(\frac{1}{\cos 4\alpha}-\frac{1}{\cos 2\alpha}\right)r^3\cos 3\theta\right]_{x=a\cos\alpha} \\ &\quad +\frac{q_0}{D}\sum_{m=1}^{M} [b_m\{(\lambda_m+2)\cos\lambda_m\theta+\cos(\lambda_m+2)\theta\}r^{\lambda_m+2}]_{x=a\cos\alpha}. \end{aligned} \tag{3.102}$$

This equation is again satisfied approximately by the method of least squares which requires

$$\delta\int_{-a\sin\alpha}^{a\sin\alpha}\left[\frac{\partial w}{\partial x}\right]^2_{x=a\cos\alpha} dy = 0 \tag{3.103}$$

and this leads to another system of M linear simultaneous equations

$$\left.\begin{aligned} A'_{11}a_1 + A'_{12}a_2 + A'_{13}a_3 + \cdots + A'_{1M}a_M &= A'_1 \\ A'_{21}a_1 + A'_{22}a_2 + A'_{23}a_3 + \cdots + A'_{2M}a_M &= A'_2 \\ A'_{31}a_1 + A'_{32}a_2 + A'_{33}a_3 + \cdots + A'_{3M}a_M &= A'_3 \\ \cdots\cdots\cdots\cdots\cdots\cdots\cdots\cdots \\ A'_{M1}a_1 + A'_{M2}a_2 + A'_{M3}a_3 + \cdots + A'_{MM}a_M &= A'_M. \end{aligned}\right\} \tag{3.104}$$

The A'_{mn} and A'_m are given by

$$A'_{mn} = A'_{nm} = \frac{A_{mn}}{16(\lambda_m + 2)(\lambda_n + 2)}, \tag{3.105}$$

where the A_{mn} are given by (3.100), and

$$A'_m = (\lambda_m + 1)\int_{-a\sin\alpha}^{a\sin\alpha} [\chi r^{\lambda_m} \cos\lambda_m\theta]_{x=a\cos\alpha}\,dy,$$

where

$$\chi = \frac{1}{16}\left[\left(1 - \frac{1}{\cos 2\alpha}\right) r^3 \cos\theta + \frac{1}{3}\left(\frac{1}{\cos 4\alpha} - \frac{1}{\cos 2\alpha}\right) r^3 \cos 3\theta\right]$$
$$+ \sum_{m=1}^{M} b_m\{(\lambda_m + 2)\cos\lambda_m\theta + \cos(\lambda_m + 2)\theta\} r^{\lambda_m + 2}. \tag{3.106}$$

Numerical results

Morley has derived numerical results for several values of the angle α, while the particular value $\alpha = 75°$ has been examined in detail as it represents a large degree of skewness. For this case, the moments M_r and M_θ along the half diagonal OC have been calculated for two levels of approximation, namely by taking $M = 3$ and 8, and the results shown in Fig. 3.12 indicate a very satisfactory degree of convergence. In particular, the term a_1, which governs the singular behaviour, changes by only 1.7 per cent between these two values of M. For $M = 8$, the absolute accuracy may be gauged from the fact that the maximum error in the boundary conditions (3.93) or (3.94) is less than 0.15 per cent.

3.10 Multi-layered plate with coupling between moments and planar strains

We conclude with a solution for a uniformly loaded clamped elliptical multi-layered plate with general coupling between moments and planar strains. This relatively simple solution is of value in estimating the effect of such coupling in more complicated cases. It is convenient to work in terms of displacements (see Section 1.8.6), so that the boundary conditions

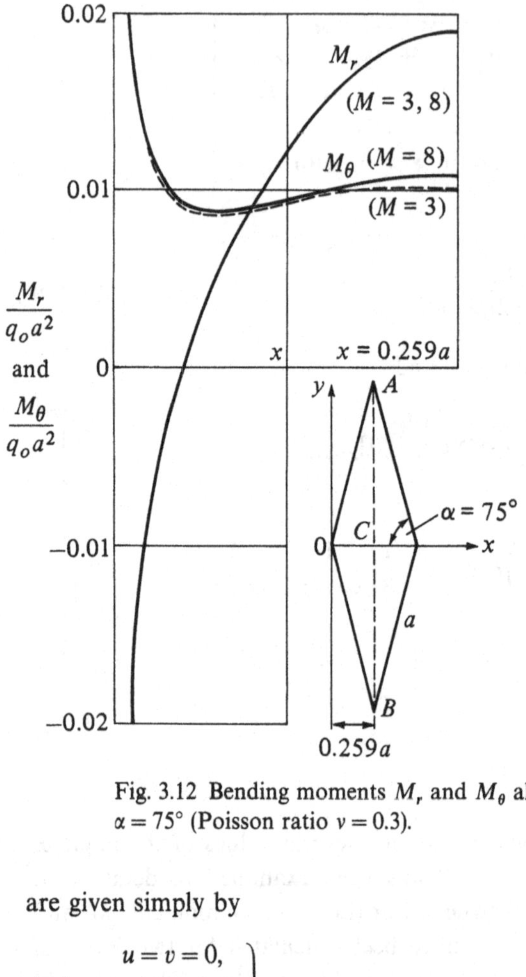

Fig. 3.12 Bending moments M_r and M_θ along the half diagonal OC for $\alpha = 75°$ (Poisson ratio $\nu = 0.3$).

are given simply by

$$\left.\begin{aligned} &u = v = 0, \\ \text{and} \quad & \\ &w = \frac{\partial w}{\partial \boldsymbol{n}} = 0. \end{aligned}\right\} \tag{3.107}$$

We now search for a solution that satisfies (3.107) in the form:

$$\left.\begin{aligned} u &= w_0(\lambda_1 x + \mu_1 y)\left(1 - \frac{x^2}{a^2} - \frac{y^2}{b^2}\right), \\ v &= w_0(\lambda_2 x + \mu_2 y)\left(1 - \frac{x^2}{a^2} - \frac{y^2}{b^2}\right), \end{aligned}\right\} \tag{3.108}$$

where $w_0, \lambda_1, \mu_1, \lambda_2, \mu_2$ are constants to be determined, and

$$w = w_0\left(1 - \frac{x^2}{a^2} - \frac{y^2}{b^2}\right)^2. \tag{3.109}$$

Substitution of (3.108), (3.109) in (1.111) and (1.112) yields two equations in which each term varies linearly with x or y. For these equations to be satisfied it is necessary that the coefficients of the terms that vary with x or y vanish and this gives rise to the following four equations that enable us to express the constants λ_i, μ_i in terms of ζ $(=a/b)$ and the elastic properties of the plate:

$$\begin{bmatrix} 3A_{11}+\zeta^2A_{66}, & 2A_{16}, & 3A_{16}+\zeta^2A_{26}, & A_{12}+A_{66} \\ 2\zeta^2A_{16}, & A_{11}+3\zeta^2A_{66}, & \zeta^2(A_{12}+A_{66}), & A_{16}+3\zeta^2A_{26} \\ 3A_{16}+\zeta^2A_{26}, & A_{12}+A_{66}, & 3A_{66}+\zeta^2A_{22}, & 2A_{26} \\ \zeta^2(A_{12}+A_{66}), & A_{16}+3\zeta^2A_{26}, & 2\zeta^2A_{26}, & A_{66}+3\zeta^2A_{22} \end{bmatrix} \cdot \begin{bmatrix} \lambda_1 \\ \mu_1 \\ \lambda_2 \\ \mu_2 \end{bmatrix}$$

$$= -\frac{4}{a^2}\begin{bmatrix} 3B_{11}+\zeta^2(B_{12}+2B_{66}) \\ 3\zeta^2(B_{16}+\zeta^2B_{26}) \\ 3(B_{16}+\zeta^2B_{26}) \\ \zeta^2(3\zeta^2B_{22}+B_{12}+2B_{66}) \end{bmatrix}. \tag{3.110}$$

Finally, substitution of (3.108), (3.109) into (1.114) gives

$$w_0 = \frac{a^4 q_0}{K}, \tag{3.111}$$

where

$$K = 8\{3D_{11} + 2\zeta^2(D_{12}+2D_{66}) + 3\zeta^4 D_{22}\}$$
$$+ 2a^2[3\{\lambda_1 B_{11} + \zeta^2\mu_2 B_{22} + (\lambda_2+\mu_1)(B_{16}+\zeta^2 B_{26})\}$$
$$+ (\zeta^2\lambda_1 + \mu_2)(B_{12}+2B_{66})].$$

References

Carrier, G. F. The bending of the clamped sectorial plate. *J. Appl. Mech.*, **66**, p. A-134 (1944).

Diaz, G. B., and Greenberg, H. G. Upper and lower bounds for the solution of the first biharmonic boundary value problem. *J. Math. Phys.*, **27**, p. 193 (1948).

Flügge, W. *Baulingenieur*, **10**, p. 221 (1929).

Morley, L. S. D. Bending of a simply supported rhombic plate under uniform normal loading. *Quart. J. Mech. Appl. Math.*, **15**, pp. 413–26 (1962).

———. *Skew plates and structures.* Pergamon Press, 1963.

Poisson, S. D. *Memoirs of the Academy*, 8. Paris, 1829.

Timoshenko, S. P., and Woinowsky-Krieger, S. *Theory of plates and shells.* 2nd ed. Chaps. 3, 9. McGraw-Hill, 1959.

Whittaker, E. T., and Watson, G. N. *A course of modern analysis.* 4th ed. Cambridge, 1940.

Williams, M. L. Surface stress singularities resulting from various boundary conditions in angular corners of plates under bending. U.S. Nat. Cong. Appl. Mech. Illinois Inst. Tech., Chicago, 1951.

Woinowsky-Krieger. *Ingenieur-Archiv.*, **4**, p. 254 (1933).

4

Plates whose boundaries are amenable to conformal transformation

Since about 1910 the Russian school of elasticians has developed powerful and elegant methods of analysis of the biharmonic equation. Prominent in this field is Muskhelishvili (1933), who gave particular attention to the boundary value problems of plane stress. The analogous boundary value problems for plates have also been considered by such authors as Lourie (1928), Lekhnitzky (1938), Vekua (1942) and Fridman (1952). Parallel with this has been work in English by authors including Stevenson (1942), Green and Zerna (1954), and Jones (1957). An essential ingredient in this method of analysis is the use of complex coordinates, and with this in view we consider first some definitions and notations.

(i) Let a be a complex number equal to $(\alpha + i\beta)$ where α and β are real; the conjugate complex number $(\alpha - i\beta)$ is then denoted by $\bar{a}$. We shall here be using complex coordinates $(z, \bar{z})$ where

$$\left.\begin{aligned} z &= x + iy, \\ \bar{z} &= x - iy. \end{aligned}\right\} \tag{4.1}$$

(ii) Let $f(z)$ be a complex polynomial function of z given by

$$f(z) = \sum_n a_n z^n = \sum_n (\alpha_n + i\beta_n) z^n.$$

The following notation is then employed:

$$\bar{f}(z) = \sum_n \bar{a}_n z^n = \sum_n (\alpha_n - i\beta_n) z^n.$$

Thus, if

$$f(z) = f_1(x, y) + if_2(x, y),$$

where f_1, f_2 are real functions, it follows that

$$\bar{f}(\bar{z}) = f_1(x, y) - if_2(x, y).$$

We will also have occasion to use the symbols $\mathscr{R}$ for 'real part of' and $\mathscr{T}$ for 'imaginary part of'. Finally, a prime will be used to denote differentiation of a function of a single complex variable.

4.1 Governing differential equation in complex coordinates

In virtue of (4.1) we have

$$\left.\begin{aligned}
\frac{\partial}{\partial x} &= \left(\frac{\partial z}{\partial x}\right)_y \frac{\partial}{\partial z} + \left(\frac{\partial \bar{z}}{\partial x}\right)_y \frac{\partial}{\partial \bar{z}} \\
&= \frac{\partial}{\partial z} + \frac{\partial}{\partial \bar{z}}
\end{aligned}\right.$$

and similarly,

$$\frac{\partial}{\partial y} = i\left(\frac{\partial}{\partial z} - \frac{\partial}{\partial \bar{z}}\right), \tag{4.2}$$

so that

$$\left.\begin{aligned}
\frac{\partial}{\partial z} &= \frac{1}{2}\left(\frac{\partial}{\partial x} - i\frac{\partial}{\partial y}\right), \\
\frac{\partial}{\partial \bar{z}} &= \frac{1}{2}\left(\frac{\partial}{\partial x} + i\frac{\partial}{\partial y}\right)
\end{aligned}\right\}$$

and therefore

$$\nabla^2 = \frac{\partial^2}{\partial x^2} + \frac{\partial^2}{\partial y^2} = 4\frac{\partial^2}{\partial z \partial \bar{z}}. \tag{4.3}$$

The governing differential equation (1.29) for plates of constant rigidity now assumes the form

$$\frac{\partial^4 w}{\partial z^2 \partial \bar{z}^2} = \frac{q}{16D} \tag{4.4}$$

and in discussing the solution of (4.4) it is convenient to write

$$w = w_p + w_c, \tag{4.5}$$

where w_p is a particular integral satisfying (4.4) and w_c is the complementary function satisfying the equation

$$\frac{\partial^4 w_c}{\partial z^2 \partial \bar{z}^2} = 0. \tag{4.6}$$

4.1.1 Particular integrals

A particular integral of (4.4) is readily found by repeated integration:

$$w_p = \frac{1}{16D}\iiiint q \, \mathrm{d}z \, \mathrm{d}z \, \mathrm{d}\bar{z} \, \mathrm{d}\bar{z}. \tag{4.7}$$

This form is particularly useful whenever q can be expressed as a polynomial in x, y for it may then also be expressed as a polynomial in $z, \bar{z}$ in virtue of (4.1). The lower limits of integration in (4.7) are arbitrary, but it is assumed that they are chosen to ensure that w_p is real; in many practical cases, lower limits of zero will suffice. Thus when $q = q_0$, a constant, we may take

$$w_p = \frac{q_0}{64D} z^2 \bar{z}^2. \tag{4.8}$$

If the load is concentrated at a point, a particular integral is more conveniently found from the results of Section 3.1.1. Thus, if there is a concentrated downward load P acting at the origin, we find from (3.9), (3.12) and (3.18) that

$$w_p = \frac{P}{8\pi D} r^2 \ln r$$

$$= \frac{P}{16\pi D} z\bar{z} \ln (z\bar{z}). \tag{4.9}$$

If the load P acts at the point $(z_0, \bar{z}_0)$ a particular integral is therefore given by

$$w_p = \frac{P}{16\pi D} (z - z_0)(\bar{z} - \bar{z}_0) \ln \{(z - z_0)(\bar{z} - \bar{z}_0)\}. \tag{4.10}$$

4.1.2 Form of the complementary function

Equation (4.6) may be integrated immediately to give

$$w_c = \bar{z}\varphi(z) + z\varphi_0(\bar{z}) + \chi(z) + \chi_0(\bar{z}).$$

where $\varphi, \varphi_0, \chi, \chi_0$ are arbitrary analytic functions; but this expression is too general because w_c is necessarily real, and accordingly we must take

$$w_c = \bar{z}\varphi(z) + z\bar{\varphi}(\bar{z}) + \chi(z) + \bar{\chi}(\bar{z}). \tag{4.11}$$

The arbitrary functions φ and χ are referred to as the *complex potentials*. Now it may be verified that w_0 is unaltered if we replace

$$\varphi(z) \quad \text{by} \quad \varphi(z) + A + iB + iCz$$

and

$$\chi(z) \quad \text{by} \quad \chi(z) - (A - iB)z + iD,$$

where A, B, C, D are arbitrary real constants, and it follows that the complex potentials φ and χ cannot be uniquely defined unless some restrictions are imposed upon them. Here, uniqueness of the functions φ and χ is achieved by adopting the convention that

$$\left.\begin{aligned} \varphi(0) &= 0, \\ \mathscr{T}\varphi'(0) &= 0, \\ \text{and} \qquad \mathscr{T}\chi(0) &= 0. \end{aligned}\right\} \tag{4.12}$$

4.1.3 Boundary conditions in terms of complex coordinates $z, \bar{z}$

The functions φ and χ are to be determined from the boundary conditions which must first be expressed in terms of the complex coordinates $z, \bar{z}$. In what follows, attention will be devoted to the clamped and simply supported cases.

Clamped boundary condition. The clamped boundary condition is given by (1.67) and (1.68). Apart from an unspecified rigid body displacement, the vanishing of w is equivalent to the vanishing of $\partial w/\partial s$, and accordingly (1.67) and (1.68) may be combined to yield the following single equation:

$$\begin{aligned} 0 &= \left(\frac{\partial w}{\partial \boldsymbol{n}}\right)^2 + \left(\frac{\partial w}{\partial s}\right)^2 \\ &= \left(\frac{\partial w}{\partial x}\right)^2 + \left(\frac{\partial w}{\partial y}\right)^2 \\ &= 4\frac{\partial w}{\partial z}\frac{\partial w}{\partial \bar{z}} \text{ in virtue of (4.2).} \end{aligned} \tag{4.13}$$

Now $\partial w/\partial z$ and $\partial w/\partial \bar{z}$ are conjugate complex quantities and the vanishing of one implies the vanishing of the other, so no generality is lost by writing (4.13) in the following simple and convenient form

$$\frac{\partial w}{\partial \bar{z}} = 0. \tag{4.14}$$

Simply supported boundary condition. The simply supported boundary condition is given by (1.69) and (1.70), and the latter may be rearranged using the identities (1.66), to give

$$\begin{aligned} \frac{\partial^2 w}{\partial \boldsymbol{n}^2} + \nu\frac{\partial^2 w}{\partial \boldsymbol{t}^2} &= \nabla^2 w - (1-\nu)\frac{\mathrm{d}\psi}{\mathrm{d}s}\frac{\partial w}{\partial \boldsymbol{n}} \\ &= 0. \end{aligned} \tag{4.15}$$

Equation (4.15) must now be expressed in terms of the complex coordinates, and to achieve this it is noted that along the boundary (see Fig. 1.6)

$$2\frac{\partial w}{\partial \bar{z}} = \frac{\partial w}{\partial x} + i\frac{\partial w}{\partial y}$$

$$= (\sin\psi - i\cos\psi)\frac{\partial w}{\partial \boldsymbol{n}}$$

$$= -ie^{i\varphi}\frac{\partial w}{\partial \boldsymbol{n}}. \tag{4.16}$$

Also, along the boundary,

$$\frac{\mathrm{d}z}{\mathrm{d}s} = \frac{\mathrm{d}x}{\mathrm{d}s} + i\frac{\mathrm{d}y}{\mathrm{d}s} = \cos\psi + i\sin\psi = e^{i\varphi}$$

so that

$$\left.\begin{aligned} &\frac{\mathrm{d}\bar{z}}{\mathrm{d}s} = e^{-i\varphi} \\ \text{and} \\ &\frac{\mathrm{d}\bar{z}}{\mathrm{d}z} = e^{-2i\varphi}. \end{aligned}\right\} \tag{4.17}$$

Thus

$$\psi = \tfrac{1}{2}i\ln\frac{\mathrm{d}\bar{z}}{\mathrm{d}z}$$

and

$$\frac{\mathrm{d}\psi}{\mathrm{d}s} = \tfrac{1}{2}i\frac{\mathrm{d}z}{\mathrm{d}\bar{z}}\left\{\frac{\mathrm{d}}{\mathrm{d}\bar{z}}\left(\frac{\mathrm{d}\bar{z}}{\mathrm{d}z}\right)\right\}\frac{\mathrm{d}\bar{z}}{\mathrm{d}s} = \tfrac{1}{2}ie^{i\varphi}\frac{\mathrm{d}}{\mathrm{d}\bar{z}}\left(\frac{\mathrm{d}\bar{z}}{\mathrm{d}z}\right). \tag{4.18}$$

Equation (4.15) may now be written in terms of the complex coordinates by using (4.3), (4.16) and (4.18):

$$4\frac{\partial^2 w}{\partial z\,\partial \bar{z}} + (1-\nu)\left\{\frac{\mathrm{d}}{\mathrm{d}\bar{z}}\left(\frac{\mathrm{d}\bar{z}}{\mathrm{d}z}\right)\right\}\frac{\partial w}{\partial \bar{z}} = 0. \tag{4.19}$$

Boundary equations for the complex potentials φ, χ. The complex potentials φ, χ are to be determined from the equations formed by substitution of (4.5) and (4.11) into the appropriate boundary conditions. Thus, for the clamped plate, substitution into (4.14) yields the following equation:

$$\varphi(z) + z\bar{\varphi}'(\bar{z}) + \bar{\chi}'(\bar{z}) = -\frac{\partial w_p}{\partial \bar{z}}. \tag{4.20}$$

Similarly, the vanishing of w along the boundary yields the equation

$$\bar{z}\varphi(z) + z\bar{\varphi}(\bar{z}) + \chi(z) + \bar{\chi}(\bar{z}) = -w_p \tag{4.21}$$

and the simply supported condition (4.19) assumes the form

$$4\{\varphi'(z)+\bar{\varphi}'(\bar{z})\}+(1-\nu)\left\{\frac{\mathrm{d}}{\mathrm{d}\bar{z}}\left(\frac{\mathrm{d}\bar{z}}{\mathrm{d}z}\right)\right\}\{\varphi(z)+z\bar{\varphi}'(\bar{z})+\bar{\chi}'(\bar{z})\}$$
$$=-4\frac{\partial^2 w_p}{\partial z\,\partial\bar{z}}-(1-\nu)\left\{\frac{\mathrm{d}}{\mathrm{d}\bar{z}}\left(\frac{\mathrm{d}\bar{z}}{\mathrm{d}z}\right)\right\}\frac{\partial w_p}{\partial\bar{z}}. \tag{4.22}$$

4.1.4 Boundary conditions in terms of new complex coordinates $\zeta, \bar{\zeta}$

Equations (4.20), (4.21) and (4.22), are, of course, only valid on the boundary where x, y and hence $z, \bar{z}$ are known; but the form of these complex coordinates on the boundary does not readily lend itself to further analysis. The next step is therefore to introduce new coordinates ξ, η by means of a suitable conformal mapping function such that the region occupied by the plate in the x, y-plane becomes the region enclosed by the circle of unit radius in the ξ, η-plane. Such a transformation is always possible, and it brings with it the advantage that as a point in the x, y-plane traces out the boundary of the plate the new coordinates ξ, η trace out the unit circle; the unit circle is given by the parametric equation

$$\xi=\cos\vartheta$$
$$\eta=\sin\vartheta,$$

so that on the boundary the complex coordinates $z, \bar{z}$ transform into $\zeta, \bar{\zeta}$ where

$$\left.\begin{aligned}\zeta&=\xi+i\eta=e^{i\vartheta}=\sigma,\ \text{say,}\\ \text{and}\quad \bar{\zeta}&=\xi-i\eta=e^{-i\vartheta}=1/\sigma.\end{aligned}\right\} \tag{4.23}$$

The boundary equations may therefore be expressed in terms of a single complex coordinate σ, and this fact facilitates their solution. Thus, if the mapping function is formally represented by the relation

$$\left.\begin{aligned}z&=\omega(\zeta)\\ \text{we have}\quad \bar{z}&=\bar{\omega}(\bar{\zeta})\end{aligned}\right\} \tag{4.24}$$

and

$$\varphi(z)=\varphi\{\omega(\zeta)\}=\varphi_1(\zeta),\ \text{say,}$$

and similarly

$$\bar{\varphi}(\bar{z})=\bar{\varphi}_1(\bar{\zeta}),\quad \chi(z)=\chi_1(\zeta),\quad \bar{\chi}(\bar{z})=\bar{\chi}_1(\bar{\zeta}).$$

In addition,

$$\left.\begin{aligned}\varphi'(z) &= \frac{\partial \varphi_1(\zeta)}{\partial \zeta}\frac{\mathrm{d}\zeta}{\mathrm{d}z} = \frac{\varphi_1'(\zeta)}{\omega'(\zeta)}\\ &\text{and similarly,}\\ \bar{\varphi}'(\bar{z}) &= \frac{\bar{\varphi}_1'(\bar{\zeta})}{\bar{\omega}'(\bar{\zeta})}.\end{aligned}\right\} \tag{4.25}$$

It is also convenient to represent $\bar{\chi}'(\bar{z})$ by $\bar{\Psi}(\bar{z})$, so that

$$\bar{\chi}'(\bar{z}) = \bar{\Psi}\{\bar{\omega}(\bar{\zeta})\} = \bar{\Psi}_1(\bar{\zeta}), \text{ say.}$$

Furthermore, along the boundary

$$-\frac{\partial w_p}{\partial \bar{z}} = f(z, \bar{z}), \text{ say}$$

$$= f\{\omega(\sigma), \bar{\omega}(1/\sigma)\} = F(\sigma), \text{ a known function of } \sigma, \tag{4.26}$$

and

$$\begin{aligned}-w_p &= -w_p(z, \bar{z})\\ &= -w_p\{\omega(\sigma), \bar{\omega}(1/\sigma)\}\\ &= H(\sigma), \text{ a known function of } \sigma.\end{aligned} \tag{4.27}$$

Also

$$\frac{\mathrm{d}}{\mathrm{d}\bar{z}}\left(\frac{\mathrm{d}\bar{z}}{\mathrm{d}z}\right) = \frac{\mathrm{d}}{\mathrm{d}\bar{z}}\left(\frac{\mathrm{d}\bar{\omega}(\bar{\sigma})}{\mathrm{d}\bar{\sigma}}\frac{\mathrm{d}\sigma}{\mathrm{d}\omega(\sigma)}\frac{\mathrm{d}\bar{\sigma}}{\mathrm{d}\sigma}\right) = -\frac{\mathrm{d}}{\mathrm{d}\bar{z}}\left(\frac{\bar{\omega}'(\bar{\sigma})}{\sigma^2\omega'(\sigma)}\right)$$

$$= -\left\{\frac{\mathrm{d}}{\mathrm{d}\sigma}\left(\frac{\bar{\omega}'(\bar{\sigma})}{\sigma^2\omega'(\sigma)}\right)\right\}\frac{\mathrm{d}\sigma}{\mathrm{d}\bar{\sigma}}\frac{\mathrm{d}\bar{\sigma}}{\mathrm{d}z} = \frac{\sigma^2}{\bar{\omega}'(\bar{\sigma})}\left\{\frac{\mathrm{d}}{\mathrm{d}\sigma}\left(\frac{\bar{\omega}'(\bar{\sigma})}{\sigma^2\omega'(\sigma)}\right)\right\}$$

$$= L(\sigma), \text{ a known function of } \sigma, \tag{4.28}$$

so that, referring to (4.22), we may therefore write

$$-4\frac{\partial^2 w_p}{\partial z\,\partial \bar{z}} - (1-\nu)\left\{\frac{\mathrm{d}}{\mathrm{d}\bar{z}}\left(\frac{\mathrm{d}\bar{z}}{\mathrm{d}z}\right)\right\}\frac{\partial w_p}{\partial \bar{z}} = S(\sigma), \text{ a known function of } \sigma. \tag{4.29}$$

The boundary equations. Substitution of (4.23)–(4.29) into the boundary equations (4.20), (4.21) and (4.22) yields, respectively,

$$\left.\begin{aligned}&\varphi_1(\sigma) + J(\sigma)\cdot\bar{\varphi}_1'(1/\sigma) + \bar{\Psi}_1(1/\sigma) = F(\sigma)\\ &\text{where}\\ &J(\sigma) = \frac{\omega(\sigma)}{\bar{\omega}'(1/\sigma)}, \text{ a known function of } \sigma,\end{aligned}\right\} \tag{4.30}$$

and

$$\bar{\omega}(1/\sigma)\cdot\varphi_1(\sigma)+\omega(\sigma)\cdot\bar{\varphi}_1(1/\sigma)+\chi_1(\sigma)+\bar{\chi}_1(1/\sigma)=H(\sigma). \tag{4.31}$$

Finally, on dividing throughout by $L(\sigma)$, (4.22) reduces to

$$\frac{4}{L(\sigma)}\left\{\frac{\varphi_1'(\sigma)}{\omega'(\sigma)}+\frac{\bar{\varphi}_1'(1/\sigma)}{\bar{\omega}'(1/\sigma)}\right\}$$
$$+(1-\nu)\{\varphi_1(\sigma)+J(\sigma)\bar{\varphi}_1'(1/\sigma)+\bar{\Psi}_1(1/\sigma)\}=\frac{S(\sigma)}{L(\sigma)}. \tag{4.32}$$

These three equations are the boundary equations for determining the complex potentials φ_1 and χ_1; for a clamped plate we require them to satisfy (4.30) and (4.31), while for a simply supported plate we require them to satisfy (4.31) and (4.32).

4.1.5 Form of the mapping function and complex potentials φ_1, χ_1

In what follows it is assumed that the mapping function can be expressed as a polynomial with, in general, complex coefficients c_n:

$$\left.\begin{aligned}\omega(\zeta)&=\sum_{n=1}^{N}c_n\zeta^n\\ \text{so that}\quad&\\ \bar{\omega}(\bar{\zeta})&=\sum_{n=1}^{N}\bar{c}_n\bar{\zeta}^n.\end{aligned}\right\} \tag{4.33}$$

There is no need to include a coefficient c_0 in (4.33) because this simply corresponds to a change of origin in the x, y-plane; by the same token no generality is lost in assuming that the coefficient c_1 is real, for this may always be achieved by a suitable rotation about the origin in the x, y-plane. With these restrictions on the form of the mapping function, it may be verified that the restrictions imposed in (4.12) on φ and χ to ensure their uniqueness, correspond to the following restrictions on φ_1 and χ_1:

$$\left.\begin{aligned}\varphi_1(0)&=0,\\ \mathscr{T}\varphi_1'(0)&=0\\ \text{and}\quad&\\ \mathscr{T}\chi_1(0)&=0.\end{aligned}\right\} \tag{4.34}$$

The functions φ_1 and χ_1 are analytic and single-valued within and on the unit circle, and we may therefore write

$$\varphi_1(\zeta)=\sum_{n=1}^{\infty}a_n\zeta^n,\text{ where } a_1 \text{ is real,}$$

$$\left.\begin{aligned}\chi_1(\zeta) &= \sum_{n=0}^{\infty} b_n\zeta^n, \text{ where } b_0 \text{ is real,}\\ \text{and}\qquad&\\ \Psi_1(\zeta) &= \sum_{n=0}^{\infty} e_n\zeta^n.\end{aligned}\right\} \tag{4.35}$$

The coefficients a_n, b_n (which are, in general, complex) are to be determined from the appropriate boundary conditions; the function Ψ_1 and the coefficients e_n are introduced only for convenience and will not be determined.

4.16 *Deflexion and moments in terms of complex coordinates $\zeta, \bar{\zeta}$*

Once the complex potentials $\varphi_1(\zeta)$ and $\chi_1(\zeta)$ are known, the deflexion is given by

$$w = \bar{w}(\bar{\zeta})\cdot\varphi_1(\zeta) + \omega(\zeta)\cdot\bar{\varphi}_1(\bar{\zeta}) + \chi_1(\zeta) + \bar{\chi}_1(\bar{\zeta}) + w_p. \tag{4.36}$$

The moments per unit length, M_x, M_y, M_{xy}, are given by (1.5), and accordingly we require expressions for the curvatures $\partial^2 w/\partial x^2, \partial^2 w/\partial y^2$, $\partial^2 w/\partial x\,\partial y$ in terms of the complex coordinates $\zeta, \bar{\zeta}$. The particular integral w_p presents no difficulty for it can be readily expressed in term of x, y. Confining attention therefore to the term w_c we note from (4.2) that

$$\frac{\partial^2 w_c}{\partial x^2} = \left(\frac{\partial^2}{\partial z^2} + 2\frac{\partial^2}{\partial z\,\partial \bar{z}} + \frac{\partial^2}{\partial \bar{z}^2}\right) w_c$$

$$\left.\begin{aligned} &= 2\mathscr{R}\{2\varphi'(z) + \bar{z}\varphi''(z) + \chi''(z)\}, \text{ from (4.11)}\\ &\text{and similarly}\\ \frac{\partial^2 w_c}{\partial y^2} &= 2\mathscr{R}\{2\varphi'(z) - \bar{z}\varphi''(z) - \chi''(z)\},\\ &\text{and}\\ \frac{\partial^2 w_c}{\partial x\,\partial y} &= -2\mathscr{T}\{\bar{z}\varphi''(z) + \chi''(z)\}.\end{aligned}\right\} \tag{4.37}$$

Equation (4.37) can be expressed in terms of $\zeta, \bar{\zeta}$ in virtue of the relations

$$\left.\begin{aligned}\varphi'(z) &= \frac{\varphi_1'(\zeta)}{\omega'(\zeta)},\\ \varphi''(z) &= \frac{\omega'(\zeta)\varphi_1''(\zeta) - \omega''(\zeta)\varphi_1'(\zeta)}{\{\omega'(\zeta)\}^3}\\ &\text{and}\\ \chi''(z) &= \frac{\omega'(\zeta)\chi_1''(\zeta) - \omega''(\zeta)\chi_1'(\zeta)}{\{\omega'(\zeta)\}^3}.\end{aligned}\right\} \tag{4.38}$$

4.2 General solution for a clamped plate

4.2.1 *Determination of complex potential $\varphi_1(\zeta)$*

Here it will be shown how the complex potential φ_1 may be determined in series form from (4.30). Once this complex potential is known, a separate analysis, based on (4.31), is preferable for the determination of the complex potential χ_1. First, however, the functions $F(\sigma)$ and $J(\sigma)$ must be expressed in powers of σ:

$$F(\sigma) = \sum_{n=-\infty}^{\infty} A_n \sigma^n, \text{ say,} \tag{4.39}$$

and

$$\begin{aligned} J(\sigma) &= \frac{c_1\sigma + c_2\sigma^2 + \cdots + c_N\sigma^N}{\bar{c} + 2\bar{c}_2\bar{\sigma} + \cdots + N\bar{c}_N\bar{\sigma}^{(N-1)}} \\ &= \sigma^N\left(\frac{c_N\sigma^{N-1} + \cdots + c_2\sigma + c_1}{\bar{c}_1\sigma^{N-1} + 2\bar{c}_2\sigma^{N-2} + \cdots + N\bar{c}_N}\right) \\ &= \sum_{n=1}^{N} g_n\sigma^n + \sum_{k=0}^{\infty} g_{-k}\sigma^{-k}, \text{ say,} \end{aligned} \tag{4.40}$$

where the coefficients g_n are obtained by straightforward long division. The coefficients g_{-k} are not required in the subsequent analysis and so need not be determined. Note that for the practically important case of uniform loading the highest power of σ occurring in $F(\sigma)$ is $(2N-1)$, and this is also the highest power of ζ occurring in $\varphi_1(\zeta)$.

Substitution of (4.35), (4.39) and (4.40) in (4.30) now gives

$$\begin{aligned} &\sum_{n=1}^{\infty} a_n\sigma^n + \left(\sum_{n=1}^{N} g_n\sigma^n + \sum_{k=0}^{\infty} g_{-k}\sigma^{-k}\right)\sum_{n=1}^{\infty} n\bar{a}_n\sigma^{-(n-1)} \\ &\quad + \sum_{n=0}^{\infty} e_n\sigma^{-n} = \sum_{n=-\infty}^{\infty} A_n\sigma^n. \end{aligned} \tag{4.41}$$

By equating coefficients of the positive powers of σ we obtain the following relations:

$$\left.\begin{aligned} a_n &= A_n & &(n \geqslant N+1), \\ a_N + \bar{a}_1 g_N &= A_N & &(n = N), \\ a_{N-1} + \bar{a}_1 g_{N-1} + 2\bar{a}_2 g_N &= A_{N-1} & &(n = N-1), \\ &\cdots\cdots\cdots\cdots\cdots\cdots\cdots\cdots\cdots\cdots \\ a_1 + \bar{a}_1 g_1 + 2\bar{a}_2 g_2 + \cdots + N\bar{a}_n g_n &= A_1 & &(n = 1). \end{aligned}\right\} \tag{4.42}$$

Equation (4.42) is sufficient to determine the coefficients a_n and hence the function $\varphi_1(\zeta)$. If the coefficients g_n and A_n are real, the coefficients a_n will also be real. If any of the coefficients g_n or A_n are complex it will be necessary to write each coefficient a_n (apart from a_1) in the form $(\alpha_n + i\beta_n)$ and to equate separately the real and imaginary parts in (4.42).

4.2.2 *Determination of complex potential $\chi_1(\zeta)$*

A similar analysis, based on (4.31), may now be employed to determine the complex potential $\chi_1(\zeta)$. First, however, the function $H(\sigma)$ is expressed in powers of σ:

$$H(\sigma) = \sum_{n=-\infty}^{\infty} B_n \sigma^n, \text{ say,} \tag{4.43}$$

and it is convenient to introduce the notation

$$\begin{aligned} \bar{\omega}(1/\sigma)\cdot\varphi_1(\sigma) &= \left(\sum_{n=1}^{N} \bar{c}_n \sigma^{-n}\right)\left(\sum_{n=1}^{\infty} a_n \sigma^n\right) \\ &= \sum_{n=-(N-1)}^{\infty} K_n \sigma^n, \text{ say.} \end{aligned} \tag{4.44}$$

Substitution of (4.35), (4.43) and (4.44) in (4.31) now gives

$$\begin{aligned} &\sum_{n=0}^{\infty} b_n \sigma^n + \sum_{k=0}^{\infty} \bar{b}_k \sigma^{-k} \\ &\quad = \sum_{n=-\infty}^{\infty} B_n \sigma^n - \sum_{n=-(N-1)}^{\infty} K_n \sigma^n - \sum_{k=-(N-1)}^{\infty} \bar{K}_k \sigma^{-k} \end{aligned} \tag{4.45}$$

and by equating coefficients of the positive powers of σ in (4.45) we obtain the following relations for determining the coefficients b_n,

$$\left.\begin{aligned} b_0 + \bar{b}_0 = 2b_0 = B_0 - (K_0 + \bar{K}_0) \quad (n=0), \\ b_n = B_n - (K_n + \bar{K}_{-n}) \quad (1 \leqslant n \leqslant N-1), \\ b_n = B_n - K_n \quad (n \geqslant N). \end{aligned}\right\} \tag{4.46}$$

Both complex potentials are now known and the problem is solved, for the deflexion at any point may be determined from (4.36), and the moments from Section 4.1.6. Note that for the case of uniform loading the highest power of ζ occurring in $\chi_1(\zeta)$ is $(2N-2)$.

4.2.3 *Bending moments along the boundary*

In the majority of practical cases the maximum bending moments in the plate occur along the clamped boundary where, as Stevenson noted, the moment M_n assumes a particularly simple form. Thus, on a clamped boundary,

$$M_n = -D\left(\frac{\partial^2 w}{\partial n^2} + \nu \frac{\partial^2 w}{\partial t^2}\right)$$

$$= -D\left\{\nabla^2 w - (1-\nu)\left(\frac{\partial^2 w}{\partial s^2} + \frac{d\psi}{ds}\frac{\partial w}{\partial \boldsymbol{n}}\right)\right\}, \text{ in virtue of (1.49)}$$

$$= -D\nabla^2 w, \text{because } \frac{\partial^2 w}{\partial s^2} \text{ and } \frac{\partial w}{\partial \boldsymbol{n}} \text{ vanish on a clamped boundary,}$$

$$= -4D\{\varphi'(z) + \bar{\varphi}'(\bar{z})\} - D\nabla^2 w_p, \text{ in virtue of (4.3) and (4.11),}$$

$$= -8D\mathscr{R}\frac{\varphi_1'(\sigma)}{\omega'(\sigma)} - D\nabla^2 w_p. \tag{4.47}$$

4.3 General solution for a simply supported plate

The complex potential $\varphi_1(\zeta)$ is to be determined from (4.32) by equating coefficients of the positive powers of σ, but the form of this equation precludes the possibility of obtaining a simple general solution such as was presented for the clamped plate. However, in Section 4.3.1 the steps in the analysis will be briefly outlined for the practically important case of uniform loading. It is to be noted that once the complex potential $\varphi_1(\zeta)$ is known, the determination of $\chi_1(\zeta)$ from (4.31) is precisely the same as that given in Section 4.2.2.

4.3.1 Uniformly loaded plate

It may be verified by substitution of (4.33) into (4.28) that $1/L(\sigma)$ may be expressed in the form

$$\frac{1}{L(\sigma)} = \sum_{n=1}^{N} l_n \sigma^n + \sum_{k=0}^{\infty} l_{-k}\sigma^{-k}$$

and for the uniformly loaded plate, in which w_p is given by (4.8), we have, therefore,

$$\frac{S(\sigma)}{L(\sigma)} = -\frac{q_0}{32D}\omega(\sigma)\cdot\bar{\omega}(\bar{\sigma})\left(\frac{8}{L(\sigma)} + (1-\nu)\omega(\sigma)\right)$$

$$= \sum_{n-1}^{2N-1} s_n \sigma^n + \sum_{k=0}^{\infty} s_{-k}\sigma^{-k}, \text{ say,} \tag{4.48}$$

where, as will be seen later, the coefficients s_{-k} are not required for the subsequent analysis.

Similarly we may write

$$\frac{4}{L(\sigma)\omega'(\sigma)} = u_1\sigma + \sum_{m=0}^{2N-2} u_{-m}\sigma^m + \sum_{k=2N-1}^{\infty} u_{-k}\sigma^{-k} \tag{4.49}$$

and

$$\frac{4}{L(\sigma)\bar{\omega}'(1/\sigma)} = \sum_{n=1}^{N} v_n\sigma^n + \sum_{k=0}^{\infty} v_{-k}\sigma^{-k} \tag{4.50}$$

where the coefficients u_{-k}, v_{-k} need not be evaluated.

Substitution of (4.35), (4.48), (4.49) and (4.50) in (4.32) now yields the following equation in which, for simplicity, terms which do not contribute to positive powers of σ are omitted:

$$\left(u_1\sigma + \sum_{m=0}^{2N-2} u_{-m}\sigma^{-m}\right)\sum_{n=1}^{\infty} na_n\sigma^{n-1} + \left(\sum_{n=1}^{N} v_n\sigma^n\right)\sum_{n=1}^{\infty} n\bar{a}_n\sigma^{-(n-1)}$$
$$+(1-\nu)\left\{\sum_{n=1}^{\infty} a_n\sigma^n + \left(\sum_{n=1}^{N} g_n\sigma^n\right)\sum_{n=1}^{\infty} n\bar{a}_n\sigma^{-(n-1)}\right\}$$
$$= \sum_{n=1}^{2N-1} s_n\sigma^n + \cdots. \tag{4.51}$$

By equating coefficients of the positive powers of σ we obtain the following relations:

$$a_n = 0 \quad (n \geqslant 2N),$$
$$u_1(2N-1)a_{2N-1} + (1-\nu)a_{2N-1} = s_{2N-1} \quad (n = 2N-1),$$
$$u_1(2N-2)a_{2N-2} + u_0(2N-1)a_{2N-1} + (1-\nu)a_{2N-2} = s_{2N-2} \quad (n = 2N-2),$$

....

$$u_1Na_N + u_0(N+1)a_{N+1} + \cdots + u_{-(N-2)}(2N-1)a_{2N-1} + (1-\nu)a_N$$
$$+ \{v_N\bar{a}_1 + (1-\nu)g_N\bar{a}_1\} = s_N \quad (n = N), \tag{4.52}$$

....

These relations suffice to determine the coefficients a_n and hence the function $\varphi_1(\zeta)$.

4.4 Square plate with rounded corners

As an illustrative example, consider the uniformly loaded and clamped plate shown in Fig. 4.1 whose boundary is given by the parametric equation

$$x = \tfrac{25}{48}a(\cos\vartheta - \tfrac{1}{25}\cos 5\vartheta),$$
$$y = \tfrac{25}{48}a(\sin\vartheta - \tfrac{1}{25}\sin 5\vartheta).$$

The appropriate mapping function is one of a family treated by Stevenson (1943), namely

$$z = \omega(\zeta) = L(\zeta + \lambda\zeta^5),$$

and is obtained from this by taking

$$L = \tfrac{25}{48}a, \quad \lambda = -\tfrac{1}{25}. \tag{4.53}$$

Now a particular integral is given by (4.8), so that along the boundary

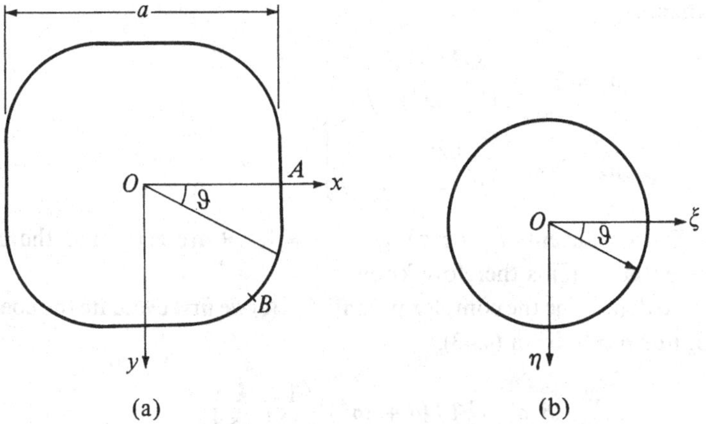

Fig. 4.1

$$\left.\begin{aligned} F(\sigma) &= -\frac{q_0}{32D} z^2 \bar{z} \\ &= T(\sigma + \lambda\sigma^5)^2 \left(\frac{1}{\sigma} + \frac{\lambda}{\sigma^5}\right), \end{aligned}\right\} \tag{4.54}$$

where

$$T = \frac{q_0 L^3}{32D},$$

and the coefficients A_n are accordingly given by the equation

$$\sum_{n=-\infty}^{\infty} A_n \sigma^n = T\{\lambda^2\sigma^9 + (2\lambda + \lambda^3)\sigma^5 + (1 + 2\lambda^2)\sigma + \lambda\sigma^{-3}\}. \tag{4.55}$$

Similarly, the coefficients g_n are given by (4.40):

$$\begin{aligned} J(\sigma) &= \sigma^5 \left(\frac{1 + \lambda\sigma^4}{\sigma^4 + 5\lambda}\right) \\ &= \lambda\sigma^5 + (1 - 5\lambda^2)\sigma + \cdots \end{aligned}$$

so that

$$g_5 = \lambda, \quad g_4 = g_3 = g_2 = 0, \quad g_1 = (1 - 5\lambda^2).$$

Equation (4.42) now gives

$$a_9 = T\lambda^2, \tag{4.56}$$

and

$$\begin{aligned} a_5 + \bar{a}_1\lambda &= T(2\lambda + \lambda^3), \\ a_1 + \bar{a}_1(1 - 5\lambda^2) + 5\bar{a}_5\lambda &= T(1 + 2\lambda^2), \end{aligned}$$

whence

$$\left.\begin{aligned} a_1 &= T\left(\frac{1-8\lambda^3-5\lambda^4}{2(1-5\lambda^2)}\right), \\ a_5 &= T\lambda\left(\frac{3-10\lambda^2-5\lambda^4}{2(1-5\lambda^2)}\right). \end{aligned}\right\} \tag{4.57}$$

The coefficients a_n for values of $n \neq 1, 5, 9$ are zero, and the complex potential $\varphi_1(\zeta)$ is therefore known.

To determine the complex potential $\chi_1(\zeta)$ we first evaluate the coefficients B_n (for $n \geqslant 0$) from (4.43),

$$\sum_{n=-\infty}^{\infty} B_n \sigma^n = \tfrac{1}{2}TL(\sigma + \lambda\sigma^5)^2\left(\frac{1}{\sigma} + \frac{\lambda}{\sigma^5}\right)^2$$

whence

$$B_0 = \tfrac{1}{2}TL(1+4\lambda^2+\lambda^4), \quad B_4 = TL\lambda(1+\lambda^2), \quad B_8 = \tfrac{1}{2}TL\lambda^2. \tag{4.58}$$

By the same token, the coefficients K_n from (4.44) are given by

$$\left.\begin{aligned} K_{-4} &= \tfrac{1}{2}TL\lambda\left(\frac{1-8\lambda^2-5\lambda^4}{1-5\lambda^2}\right), \\ K_0 &= \tfrac{1}{2}TL\left(\frac{1-5\lambda^2-15\lambda^4-5\lambda^6}{1-5\lambda^2}\right), \\ K_4 &= \tfrac{1}{2}TL\lambda\left(\frac{3-8\lambda^2-15\lambda^4}{1-5\lambda^2}\right), \quad K_8 = TL\lambda^2. \end{aligned}\right\} \tag{4.59}$$

The coefficients b_n are now given immediately from (4.46), (4.58) and (4.59):

$$\begin{aligned} b_0 &= -TL\left(\frac{1-9\lambda^2-11\lambda^4-5\lambda^6}{4(1-5\lambda^2)}\right), \\ b_4 &= -TL\lambda(1+\lambda^2), \quad b_8 = -\tfrac{1}{2}TL\lambda^2. \end{aligned} \tag{4.60}$$

The coefficients b_n for values of $n \neq 0, 4, 8$ are zero, and the complex potential $\chi_1(\zeta)$ is given by (4.35) and (4.60). The deflexion at the origin is given by

$$\begin{aligned} w_0 &= 2\mathscr{R}\chi_1(0) = 2b_0 \\ &= \frac{q_0 L^4}{64D}\left(\frac{1-9\lambda^2-11\lambda^4-5\lambda^6}{1-5\lambda^2}\right) \\ &\approx 0.00114\frac{q_0 a^4}{D}, \text{ for the plate specified by (4.53).} \end{aligned}$$

The moments per unit length at the origin are given by Section 4.1.6, whence

$$(M_x)_0 = (M_y)_0 = -4(1+\nu)Da_1/L$$
$$= \frac{q_0 L^2(1+\nu)(1-8\lambda^2-5\lambda^4)}{16(1-5\lambda^2)}$$
$$\approx 0.017 q_0 a^2(1+\nu).$$

From Section 4.2.3, the edge moment per unit length at A, where $\sigma = 1$, is given by

$$(M_n)_A = -\frac{8D(a_1+5a_5+9a_9)}{L(1+5\lambda)} - \frac{q_0 a^2}{16}$$
$$\approx -0.044 q_0 a^2,$$

and the edge moment per unit length at B, where $\sigma = e^{i\pi/4}$, is given by

$$(M_n)_B = -\frac{8D(a_1-5a_5+9a_9)}{L(1-5\lambda)} - \frac{q_0(OB)^2}{4}$$
$$\approx -0.027 q_0 a^2.$$

4.4.1 *Various square plates with rounded corners*

The numerical value of the parameter λ in (4.53) was chosen so that the curvature of the boundary vanished at the point A, and this automatically fixed the curvature at the point B. By introducing a further term in the mapping function, as shown in (4.61), it is possible to construct a family of such 'rounded squares' with differing curvatures at the 'corner points':

$$\omega(\zeta) = L\left\{\zeta - \left(\frac{1+k}{25}\right)\zeta^5 + \frac{k}{81}\zeta^9\right\}. \tag{4.61}$$

4.5 Anisotropic plates

Anisotropic plates with zero coupling between **N** and **M**, as discussed in Section 1.8.3, also admit solutions in terms of functions of complex variables (Lekhnitsky 1968). Thus, referring to (1.101) and (1.102), we search for complementary functions of the form $w_c(x+\mu y)$ where w_c is an arbitrary analytic function and μ is a constant. Substitution into (1.101), with q zero, yields the following characteristic equation:

$$D_{11} + 4D_{16}\mu + 2(D_{12}+2D_{66})\mu^2 + 4D_{26}\mu^3 + D_{22}\mu^4 = 0. \tag{4.62}$$

Lekhnitsky has shown that for all elastic materials, (4.62) has only complex or purely imaginary roots; these roots are denoted by $\mu_1, \bar{\mu}_1, \mu_2$ and $\bar{\mu}_2$

where a bar denotes the conjugate and

$$\mu_1 = \alpha + i\beta, \quad \mu_2 = \gamma + i\delta, \tag{4.63}$$

where $\alpha, \beta, \gamma, \delta$ are real.

It is also convenient to introduce complex variables

$$z_1 = x + \mu_1 y \quad \text{and} \quad z_2 = x + \mu_2 y, \tag{4.64}$$

so that

$$w_c = 2\mathscr{R}\{w_1(z_1) + w_2(z_2)\}, \tag{4.65}$$

or, in the special case of equal complex roots μ,

$$w_c = 2\mathscr{R}\{w_1(z_1) + \bar{z}_1 w_2(z_1)\}. \tag{4.66}$$

References

Fridman, M. M. Solution of the general problem of bending of a thin isotropic elastic plate supported along an edge. *Prikl. Mat. i Mech.*, **16**, pp. 429–36 (1952).

Green, A. E., and Zerna, W. *Theoretical elasticity*. Oxford, 1954.

Jones, P. D. Small deflection theory of flat plates using complex variables. Parts 1, 2, 3. *A.R.L. Reports SM* 252 (1957); 260 (1958); 266 (1959). Melbourne, Australia.

Lekhnitsky, S. G. On some problems related to the theory of bending of thin strips. *Prikl. Mat. i Mech.*, **2**, pp. 181–210 (1938).

———. *Anisotropic plates*. Gordon and Breach, New York, 1968.

Lourie, A. I. On the problem of the equilibrium of plates with supported edges. *Izv. Leningr. Politechn. Inst.*, **31**, pp. 305–20 (1928).

Muskhelishvili, N. I. *Some basic problems of the mathematical theory of elasticity*. Translated by J. R. M. Radok. P. Noordhoff, 1953.

Stevenson, A. C. On the equilibrium of plates. *Phil. Mag.*, **33**, pp. 639 (1942).

———. The boundary couples in thin plates. *Phil. Mag.*, **34**, pp. 105–14 (1943).

Vekua, I. N. On the bending of plates with free edges. *Soobshcheniya A.N. Gruz. S.S.R.*, **3**, pp. 641–8 (1942).

Additional references

Bassali, W. A. Transverse bending of finite and semi-infinite thin elastic plates. *Proc. Cambridge Phil. Soc.*, **53**, pp. 248–55 (January 1957).

———. Thin circular plates supported at several points along the boundary. *Proc. Cambridge Phil. Soc.*, **53**, pp. 525–35 (April 1957).

Buchwald, T., and Tiffen, R. Boundary value problems of simply supported elastic plates. *Quart. J. Mech. Appl. Math.*, **9**, pp. 289–498 (1956).

Deverall, L. I. Solution of some problems in bending of thin clamped plates by means of the method of Muskhelishvili. *J. Appl. Mech.*, **24**, pp. 295–8 (1957).

Tiffen, R. Some problems of thin clamped elastic plates. *Quart. J. Mech. Appl. Math.*, **8**, pp. 237–50 (1955).

Yu, Y-Y. Bending of isotropic thin plates by concentrated edge couples and forces. *J. Appl. Mech.*, **21**, pp. 129–39 (1954).

5

Plates with variable rigidity

Plates with variable rigidity seldom lend themselves to exact analysis and recourse must then be had to an approximate treatment. However, there are a number of cases that do admit of exact analysis (exact, within the framework of small-deflexion plate theory), and such cases are considered here. It must be emphasized throughout that the mid-surface of the plate is assumed plane.

The cases considered in Sections 5.1 and 5.2 admit of an exact analysis in virtue of the simplicity of the applied loading. Those considered in Sections 5.3–5.6 are characterized by the simplicity of the variation of the rigidity.

5.1 Flexure and torsion of a strip of variable rigidity

We consider first the pure flexure and torsion of a strip whose rigidity $D(y)$ varies, in an arbitrary manner, across the width. The case of flexure due to shear is then considered.

5.1.1 Pure flexure

It may be verified by substitution that the deflected form

$$w = -\tfrac{1}{2}\kappa(x^2 - \nu y^2) \tag{5.1}$$

satisfies (1.28) provided that q is zero and that D does not vary with x. This deflected form gives rise to moments per unit width which may be determined from (1.5):

$$\left.\begin{aligned} &M_x = \kappa D'(y) \\ &\text{where} \\ &D'(y) = E\{t(y)\}^3/12 \\ &\text{and} \\ &M_y = M_{xy} = 0. \end{aligned}\right\} \tag{5.2}$$

Further, substitution of (5.2) in (1.26) gives

$$Q_x = Q_y = 0,$$

so that the only forces acting on the plate are moments M_x per unit width, and their resultant may be equated to the applied moment $\bar{M}$. Thus, if the strip is bounded by the lines $y = 0, b$,

$$\begin{aligned}\bar{M} &= \int_0^b M_x \, \mathrm{d}y \\ &= \kappa \int_0^b D'(y) \, \mathrm{d}y. \end{aligned} \tag{5.3}$$

The ratio $\bar{M}/\kappa$ is referred to as the *flexural rigidity* of the strip.

5.1.2 *Pure torsion*

It may likewise be shown that the deflected form

$$w = -\tau xy \tag{5.4}$$

satisfies (1.28) provided that q is zero and that D does not vary with x. Substitution of (5.4) into (1.5) then gives

$$M_{xy} = \frac{\tau}{1+\nu} D'(y) \tag{5.5}$$

and

$$M_x = M_y = 0,$$

while the shears are determined from (1.26):

$$\begin{aligned} Q_x &= \frac{\partial M_{xy}}{\partial y} \\ &= \frac{\tau}{1+\nu} \frac{\partial}{\partial y} D'(y), \end{aligned} \tag{5.6}$$

$$Q_y = 0.$$

In discussing the applied loading which gives rise to the above distribution of M_{xy} and Q_x it is convenient to regard $D'(y)$ as vanishing at the boundaries $y = 0, b$. This can be done with no loss of generality, for if the rigidity is non-zero at the true boundaries we can achieve our aim by redefining the boundaries by the lines $y = -\delta,\ b + \delta$, where δ is a vanishingly small positive length. With this proviso in mind we find that

$$[M_{xy}]_{y=0,b} = 0 \tag{5.7}$$

so that, since M_y and Q_y are zero, these edges are free. Further, the total shear force acting over the section is given by

$$\int_0^b Q_x \, dy = \int_0^b \frac{\partial M_{xy}}{\partial y} dy$$

$$= 0 \tag{5.8}$$

by virtue of (5.7).

The resultant of the forces acting over the section is therefore a torque $\bar{T}$ whose magnitude is given by

$$\bar{T} = \int_0^b M_{xy} \, dy - \int_0^b y Q_x \, dy$$

$$= \int_0^b M_{xy} \, dy - \int_0^b y \frac{\partial M_{xy}}{\partial y} dy$$

which, on integrating by parts and using (5.7),

$$= 2 \int_0^b M_{xy} \, dy$$

$$= \frac{2\tau}{1+\nu} \int_0^b D'(y) \, dy \tag{5.9}$$

by virtue of (5.5).

It is to be noticed that the torque due to the vertical shears Q_x is the same as that due to 'horizontal' shears which comprise the twisting moment M_{xy}. This equality of torsional components is indeed true for a cylinder of any cross-section.

The ratio $\bar{T}/\tau$ is referred to as the *torsional rigidity* of the strip, and a comparison of (5.3) and (5.9) shows that

$$\frac{\text{torsional rigidity of strip}}{\text{flexural rigidity of strip}} = \frac{2}{1+\nu}. \tag{5.10}$$

5.1.3 *Flexure due to shear*

Consider now the deflected form

$$w = -\frac{c}{6}\{x^3 - 3\nu x (y - y_0)^2\}, \tag{5.11}$$

where c and y_0 are constants. This deflexion is such that $(\partial w/\partial y)_{y=y_0}$, vanishes for all values of x. Furthermore, this deflexion satisfies (1.28) provided that q is zero and that D does not vary with x. The moments throughout the strip are obtained by substituting (5.11) into (1.5), which gives

$$\left.\begin{aligned} M_x &= cx(1-\nu^2)D \\ M_y &= 0 \\ M_{xy} &= -c\nu(1-\nu)(y-y_0)D, \end{aligned}\right\} \tag{5.12}$$

and it should be noted that

$$\left.\begin{aligned} [M_x]_{x=0} &= 0 \\ \text{and} \qquad [M_{xy}]_{y=0,b} &= 0, \end{aligned}\right\}$$

if we regard $D(y)$ as vanishing at the edges.

The shears per unit length are given by (1.26) and (5.12), whence

$$\left.\begin{aligned} Q_x &= c(1-\nu^2)D - c\nu(1-\nu)\frac{\partial}{\partial y}\{(y-y_0)D\} \\ \text{and} \qquad Q_y &= 0. \end{aligned}\right\} \tag{5.13}$$

The total shear force $\bar{Q}$ acting over any cross-section is therefore constant, and given by

$$\begin{aligned} \bar{Q} &= \int_0^b Q_x \, \mathrm{d}y \\ &= c\int_0^b D' \, \mathrm{d}y, \end{aligned} \tag{5.14}$$

and this equation may be regarded as determining the constant c in terms of the applied shear force.

Similarly, the total torque $\bar{T}$ acting about the line $y = y_0$, say, is the same for all cross-sections, and is given by

$$\begin{aligned} \bar{T} &= \int_0^b M_{xy} \, \mathrm{d}y - \int_0^b (y-y_0)Q_x \, \mathrm{d}y \\ &= -\frac{c\nu}{1+\nu}\int_0^b (y-y_0)D' \, \mathrm{d}y \\ &\quad - c\int_0^b (y-y_0)\left\{D' - \frac{\nu}{1+\nu}\frac{\partial}{\partial y}\{(y-y_0)D'\}\right\}\mathrm{d}y \\ &= -c\left(\frac{1+3\nu}{1+\nu}\right)\int_0^b (y-y_0)D' \, \mathrm{d}y. \end{aligned} \tag{5.15}$$

The resultant $\bar{Q}$ and $\bar{T}$ of forces acting at the end of the strip, at $x = 0$, is therefore determined. It follows from Saint Venant's principle that any applied distribution of M_{xy} and Q_x which has the same resultant $\bar{Q}$ and $\bar{T}$ will deflect the strip into a form which differs from (5.11) only in the immediate neighbourhood of the loaded face. The resultant $\bar{Q}$ and $\bar{T}$ corresponds to a shear force $\bar{Q}$ alone acting at $y=\bar{y}$, say, where

$$-\bar{Q}(\bar{y}-y_0)=\bar{T},$$

whence, from (5.14) and (5.15),

$$\bar{y}-y_0=\left(\frac{1+3\nu}{1+\nu}\right)\frac{\int_0^b (y-y_0)D'\,\mathrm{d}y}{\int_0^b D'\,\mathrm{d}y}. \tag{5.16}$$

If y_0 is at the *centroid* of the cross-section, the point at $\bar{y}$ is referred to as the *flexural centre* of the cross-section. Equation (5.16) was first derived by Duncan (1932) who treated the strip as a narrow prism rather than as a plate.

5.2 Torsion and flexure of strip with chordwise temperature variation

A chordwise temperature variation in a long strip results in a constant pattern of longitudinal stresses away from the ends, and more complex stress fields near the ends. If at least one end is free, the longitudinal stresses are self-equilibrating but, nevertheless, they affect the torsional and flexural rigidities. First we show how to determine the longitudinal stresses in a strip of variable thickness $t(y)$ with a chordwise temperature variation $T(y)$, noting that account can also be taken of symmetrical variations of temperature through the thickness if $T(y)$ is defined as the average temperature, as per Section 1.6.1.

If the strip is prevented from extending longitudinally, the longitudinal stresses are given simply by

$$\left.\begin{aligned}[\sigma_x]_{u=0}&=-E\alpha T(y),\\ \sigma_y=\tau_{xy}&=0,\end{aligned}\right\} \tag{5.17}$$

where α is the coefficient of thermal expansion.

In general, these stresses have a resultant longitudinal force and moment, but if the strip is free the stresses σ_x away from the ends are of the form

$$\sigma_x=-E\alpha T(y)+c_1+c_2 y, \tag{5.18}$$

where c_1, c_2 are such that

$$\int_0^b N_x\,\mathrm{d}y=\int_0^b t(y)\sigma_x\,\mathrm{d}y=0,$$

and

$$\left.\int_0^b yN_x\,\mathrm{d}y = \int_0^b yt(y)\sigma_x\,\mathrm{d}y = 0.\right\} \tag{5.19}$$

In what follows, we assume that the distribution of longitudinal forces per unit length, N_x, is known and we determine the effect these have on the torsional and flexural rigidities. In the latter case, we confine our attention to middle-surface forces N_x that are self-equilibrating and therefore satisfy (5.19). For the case of torsion, however, the analysis is applicable to *any* chordwise distribution of N_x and therefore, for example, to the torsion of a strip under tension.

5.2.1 *Torsion of strip with longitudinal stresses*

It may be shown that (1.63), with κ_T and q zero and $D = D(y)$, is satisfied by

$$w = -\kappa_{xy}xy + Cx, \tag{5.20}$$

where κ_{xy} is the twisting curvature or twist per unit length and Cx is a rigid body rotation whose introduction is required if the forces N_x are not self-equilibrating; alternatively, the x-axis can be chosen to coincide with the resultant of the longitudinal forces N_x.

In determining the torque $\bar{T}$ that is applied to the strip it will be seen that in addition to the components due to M_{xy} and Q_x, as in Section 5.1.2, there is also a component due to the middle-surface forces per unit length N_x because in the twisted state these have components normal to the original plane of the strip. Thus we find

$$\bar{T} = \int_0^b M_{xy}\,\mathrm{d}y - \int_0^b yQ_x\,\mathrm{d}y - \int_0^b yN_x\frac{\partial w}{\partial x}\,\mathrm{d}y. \tag{5.21}$$

But there is no *resultant* out-of-plane component of the forces N_x and hence

$$\int_0^b N_x\frac{\partial w}{\partial x}\,\mathrm{d}y = 0, \tag{5.22}$$

whence, from (5.20),

$$\left.\begin{aligned} &C = \kappa_{xy}I_1/I_0, \text{ say,}\\ \text{where}\quad &\\ &I_n = \int_0^b y^n N_x\,\mathrm{d}y.\end{aligned}\right\} \tag{5.23}$$

Integration of the third integral in (5.21) now gives

$$-\int_0^b yN_x\frac{\partial w}{\partial x}\,\mathrm{d}y = \kappa_{xy}(I_2 - I_1^2/I_0), \tag{5.24}$$

and hence the torsional rigidity is given by

$$\bar{T}/\kappa_{xy} = 2(1-\nu)\int_0^b D(y)\,\mathrm{d}y + I_2 - I_1^2/I_0. \tag{5.25}$$

Note that if the x-axis is chosen to lie along the line of the resultant of the forces N_x, the constant C and the integral I_1 are zero. Thus for a strip of constant thickness subjected to a tensile load P, we find

$$\frac{\text{torsional rigidity}}{(\text{torsional rigidity})_{P=0}} = 1 + \frac{(1+\nu)Pb}{2Et^3}. \tag{5.26}$$

5.2.2 *Flexure of strip with self-equilibrating longitudinal stresses*

The deflexion of a strip with longitudinal curvature κ_x is of the form

$$w = -\tfrac{1}{2}\kappa_x x^2 + \bar{w}(y), \tag{5.27}$$

which may be substituted into (1.63) to yield the following differential equation for $\bar{w}(y)$,

$$\frac{\mathrm{d}^2}{\mathrm{d}y^2}\left\{D\left(\frac{\mathrm{d}^2\bar{w}}{\mathrm{d}y^2} - \nu\kappa_x\right)\right\} = -\kappa_x N_x. \tag{5.28}$$

Boundary conditions

The edges of the strip are free and hence from (1.76) and (1.77)

$$\left[D\left(\frac{\mathrm{d}^2\bar{w}}{\mathrm{d}y^2} - \nu\kappa_x\right)\right]_{y=0,b} = 0, \tag{5.29}$$

and

$$\left[\frac{\mathrm{d}}{\mathrm{d}y}\left\{D\left(\frac{\mathrm{d}^2\bar{w}}{\mathrm{d}y^2} - \nu\kappa_x\right)\right\}\right]_{y=0,b} = 0. \tag{5.30}$$

Determination of $\bar{w}(y)$

Equation (5.28) may now be integrated once to give

$$\frac{\mathrm{d}}{\mathrm{d}y}\left\{D\left(\frac{\mathrm{d}^2\bar{w}}{\mathrm{d}y^2} - \nu\kappa_x\right)\right\} = -\kappa_x\int_0^y N_x\,\mathrm{d}y, \tag{5.31}$$

where the limits of integration are chosen to satisfy (5.30). Similarly, a second integration yields

$$D\left(\frac{\mathrm{d}^2\bar{w}}{\mathrm{d}y^2} - \nu\kappa_x\right) = -\kappa_x\int_0^y\int_0^y N_x\,\mathrm{d}y\,\mathrm{d}y, \tag{5.32}$$

which satisfies the boundary condition (5.29). Equation (5.32) may be further integrated to determine $\bar{w}(y)$ but, as shown below, the moment applied to the strip may be determined without recourse to such integration.

The total moment acting on the strip $\bar{M}$ is the sum of that due to flexure about the mid-surface of the strip $\bar{M}_1$, say, plus that due to middle-surface forces $\bar{M}_2$, say. Thus

$$\bar{M}_1 = \int_0^b D\left(\kappa_x - \nu\frac{d^2\bar{w}}{dy^2}\right)dy, \text{ from (5.27)}$$

$$= (1-\nu^2)\kappa_x\int_0^b D\,dy + \nu\kappa_x\int_0^b\left(\int_0^y\int_0^y N_x\,dy\,dy\right)dy, \qquad (5.33)$$

from (5.32).

The expression for $\bar{M}_2$ lends itself to repeated integration by parts. Thus

$$\bar{M}_2 = \int_0^b \bar{w}N_x\,dy$$

$$= \left[\bar{w}\int_0^y N_x\,dy\right]_0^b - \left[\frac{d\bar{w}}{dy}\int_0^y\int_0^y N_x\,dy\,dy\right]_0^b$$

$$+ \int_0^b\left\{\frac{d^2\bar{w}}{dy^2}\left(\int_0^y\int_0^y N_x\,dy\,dy\right)\right\}dy. \qquad (5.34)$$

It can be shown that the first two terms above vanish because of the equilibrium conditions (5.19), while in the third term (5.32) enables us to express $d^2\bar{w}/dy^2$ in terms of known functions. Hence we obtain

$$\left.\begin{aligned} \frac{\bar{M}}{\kappa_x} &= (1-\nu^2)\int_0^b D\,dy + 2\nu\int_0^b \Phi\,dy - \int_0^b \frac{\Phi^2}{D}dy, \\ &\text{where} \\ \Phi &= \int_0^y\int_0^y N_x\,dy\,dy, \end{aligned}\right\} \qquad (5.35)$$

and it is to be noted that this definition of Φ identifies it as the force function or, to be more precise, that version of the force function whose linear terms are chosen to make Φ vanish at $y = 0, b$.

5.3 Rectangular plate with exponential variation of rigidity

Following Conway (1958) we consider the rigidity to vary according to the equation

$$D = D_0 e^{\lambda\pi y/a} \qquad (5.36)$$

where λ is a constant and the factor π/a has been introduced for convenience (Fig. 5.1). It is to be noted that such a variation in rigidity

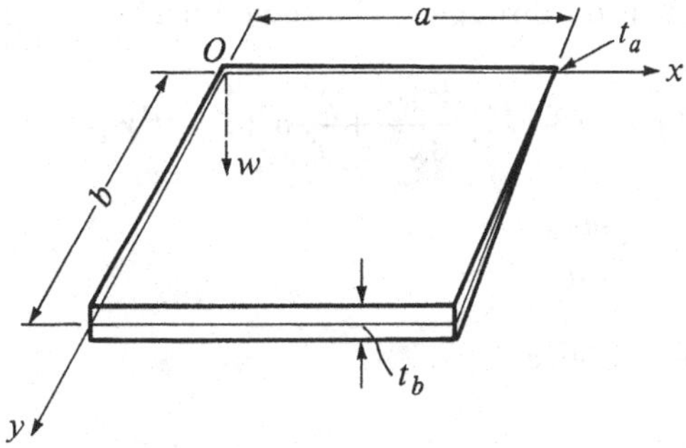

Fig. 5.1

corresponds to a thickness variation given by

$$t = t_0 e^{\lambda \pi y/3a}.$$

If the thickness varies from t_0 to t_b the coefficient λ is, therefore, given by

$$\lambda = \frac{3a}{\pi b} \ln \frac{t_b}{t_0}. \tag{5.37}$$

Substitution of (5.36) in (1.28) yields the equation

$$\nabla^4 w + \frac{2\lambda\pi}{a}\left(\frac{\partial^3 w}{\partial x^2 \partial y} + \frac{\partial^3 w}{\partial y^3}\right) + \frac{\lambda^2 \pi^2}{a^2}\left(\frac{\partial^2 w}{\partial y^2} + \nu \frac{\partial^2 w}{\partial x^2}\right) = \frac{q}{D_0} e^{-\lambda \pi y/a}. \tag{5.38}$$

In considering solutions of (5.38) attention is confined to plates simply supported along the edges $x = 0, a$ and subjected to a distributed loading of the form

$$q = e^{\beta \pi y/a} q(x) = e^{\beta \pi y/a} \sum_{m=1}^{\infty} q_m \sin \frac{m\pi x}{a}, \tag{5.39}$$

where β is an arbitrary constant.

Following the analysis of Sections 2.2 and 2.2.2 we search for a solution of (5.38) and (5.39) in the form

$$w = e^{(\beta - \lambda)\pi y/a} w_1(x) + \sum_{m=1}^{\infty} Y_m \sin \frac{m\pi x}{a}, \tag{5.40}$$

where the first term is a particular integral of (5.38) and the summation of terms the complementary integral. All terms in (5.40) are to satisfy the conditions of simple support along $x = 0, a$. Substitution of the first term in (5.40) in (5.38) and division throughout by the term $e^{(\beta-\lambda)\pi y/a}$ gives

$$\frac{d^4 w_1}{dx^4}+\frac{\pi^2}{a^2}\{2\beta(\beta-\lambda)+\nu\lambda^2\}\frac{d^2 w_1}{dx^2}+\frac{\pi^4}{a^4}\beta^2(\beta-\lambda)^2 w_1$$

$$=\frac{1}{D_0}\sum_{m=1}^{\infty} q_m \sin\frac{m\pi x}{a}, \tag{5.41}$$

which may be integrated to give

$$w_1(x)=\frac{a^4}{D_0\pi^4}\sum_{m=1}^{\infty}\frac{q_m}{(m^2+\lambda\beta-\beta^2)^2-\nu m^2\lambda^2}\sin\frac{m\pi x}{a}. \tag{5.42}$$

The differential equation for Y_m, obtained from (5.38) and (5.40), reduces to

$$a^4\frac{d^4 Y_m}{dy^4}+2\pi\lambda a^3\frac{d^3 Y_m}{dy^3}+\pi^2 a^2(\lambda^2-2m^2)\frac{d^2 Y_m}{dy^2}-2\pi^3\lambda m^2 a\frac{dY_m}{dy}$$

$$+\pi^4 m^2(m^2-\nu\lambda^2)Y_m=0, \tag{5.43}$$

the solution of which may be written in the form

$$Y_m=\frac{a^4}{D_0\pi^4}\sum_{i=1}^{4} A_{m,i}e^{r_{m,i}\pi y/a}, \tag{5.44}$$

where the $r_{m,i}$ are the roots of the equation

$$r_m^4+2\lambda r_m^3+(\lambda^2-2m^2)r_m^2-2\lambda m^2 r_m+m^2(m^2-\nu\lambda^2)=0. \tag{5.45}$$

The constants $A_{m,i}$ are to be determined from the boundary conditions along $y=0, b$. For example, the vanishing of w along these edges gives rise to the equations

$$\left.\begin{aligned}&\frac{q_m}{(m^2+\lambda\beta-\beta^2)^2-\nu m^2\lambda^2}+\sum_{i=1}^{4}A_{m,i}=0,\\&\frac{q_m e^{(\beta-\lambda)\pi b/a}}{(m^2+\lambda\beta-\beta^2)^2-\nu m^2\lambda^2}+\sum_{i=1}^{4}A_{m,i}e^{r_{m,i}\pi b/a}=0.\end{aligned}\right\} \tag{5.46}$$

Similarly, the vanishing of $\partial w/\partial y$ along $y=0, b$ gives rise to the equations

$$\left.\begin{aligned}&\frac{(\beta-\lambda)q_m}{(m^2+\lambda\beta-\beta^2)^2-\nu m^2\lambda^2}+\sum_{i=1}^{4}r_{m,i}A_{m,i}=0,\\&\frac{(\beta-\lambda)q_m e^{(\beta-\lambda)\pi b/a}}{(m^2+\lambda\beta-\beta^2)^2-\nu m^2\lambda^2}+\sum_{i=1}^{4}r_{m,i}A_{m,i}e^{r_{m,i}\pi b/a}=0,\end{aligned}\right\} \tag{5.47}$$

and the vanishing of $\partial^2 w/\partial y^2$ along $y=0, b$ gives rise to the equations

$$\left.\begin{aligned}
&\frac{(\beta-\lambda)^2 q_m}{(m^2+\lambda\beta-\beta^2)^2-\nu m^2\lambda^2}+\sum_{i=1}^{4} r_{m,i}^2 A_{m,i}=0,\\
&\frac{(\beta-\lambda)^2 q_m e^{(\beta-\lambda)\pi b/a}}{(m^2+\lambda\beta-\beta^2)^2-\nu m^2\lambda^2}+\sum_{i=1}^{4} r_{m,i}^2 A_{m,i} e^{r_{m,i}\pi b/a}=0.
\end{aligned}\right\} \tag{5.48}$$

5.3.1 *Simply supported plate under uniform load*

As an example, we consider a uniformly loaded and simply supported rectangular plate with exponentially varying rigidity. For such a plate

$$\beta=0$$

and

$$q_m=\frac{2q_0}{\pi m}\{1-(-1)^m\} \tag{5.49}$$

and the requisite boundary equations (5.46) and (5.48) reduce to

$$\left.\begin{aligned}
&\frac{2q_0\{1-(-1)^m\}}{\pi m^3(m^2-\nu\lambda^2)}+\sum_{i=1}^{4} A_{m,i}=0,\\
&\frac{2q_0\{1-(-1)^m\}e^{-\lambda\pi b/a}}{\pi m^3(m^2-\nu\lambda^2)}+\sum_{i=1}^{4} A_{m,i}e^{r_{m,i}\pi b/a}=0,\\
&\frac{2\lambda^2 q_0\{1-(-1)^m\}}{\pi m^3(m^2-\nu\lambda^2)}+\sum_{i=1}^{4} r_{m,i}^2 A_{m,i}=0,\\
&\frac{2\lambda^2 q_0\{1-(-1)^m\}e^{-\lambda\pi b/a}}{\pi m^3(m^2-\nu\lambda^2)}+\sum_{i=1}^{4} r_{m,i}^2 A_{m,i}e^{r_{m,i}\pi b/a}=0.
\end{aligned}\right\} \tag{5.50}$$

The solution of (5.50) is best obtained numerically.

5.4 Rectangular plate with linear variation of rigidity

This case was discussed by Gran Olsson (1934). The rigidity is considered to vary according to the equation

$$D=\lambda D_0 y/b. \tag{5.51}$$

If the rigidity increases from D_0 to D_1 over the width b of the plate, the coefficient λ is given by $\lambda=(D_1-D_0)/D_0$ and the origin is chosen (Fig. 5.2) so that the plate is bounded by the lines $y=y_0$, y_0+b, where

$$y_0=bD_0/(D_1-D_0).$$

Substitution of (5.51) in (1.28) yields the equation

$$\nabla^2(D\nabla^2 w)=q. \tag{5.52}$$

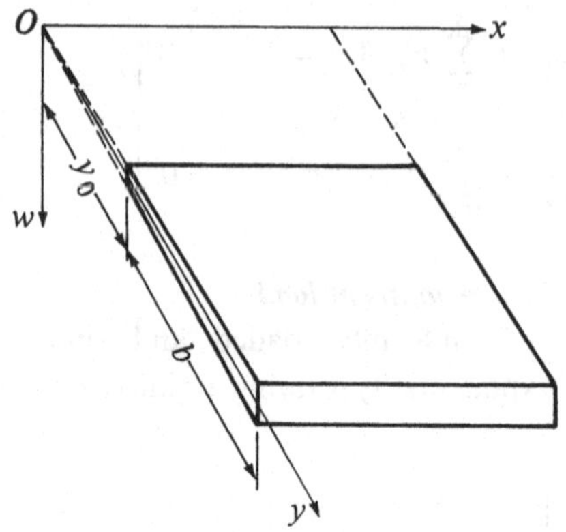

Fig. 5.2

Attention is now confined to a load distribution of the form

$$q = \frac{\lambda y}{b} q(x) = \frac{\lambda y}{b} \sum_{m=1}^{\infty} q_m \sin \frac{m\pi x}{a} \tag{5.53}$$

for which a particular integral of (5.52) is

$$w_1 = \frac{a^2}{D_0 \pi^4} \sum_{m=1}^{\infty} \frac{q_m}{m^4} \sin \frac{m\pi x}{a}. \tag{5.54}$$

We now search for a solution of (5.52) in the form

$$w = w_1 + \sum_{m=1}^{\infty} Y_m \sin \frac{m\pi x}{a}, \tag{5.55}$$

where the functions Y_m satisfy the differential equation

$$\left(\frac{d^2}{dy^2} - \frac{m^2\pi^2}{a^2}\right)\left\{y\left(\frac{d^2 Y_m}{dy^2} - \frac{m^2\pi^2}{a^2} Y_m\right)\right\} = 0. \tag{5.56}$$

The general solution of (5.56) may be expressed in terms of the *exponential integral* defined by

$$\mathrm{Ei}(u) = \int_{-\infty}^{u} \frac{e^u}{u} du, \quad \mathrm{Ei}(-u) = \int_{-\infty}^{u} \frac{e^{-u}}{u} du.$$

Introducing $\zeta = \pi y/a$, we then have

$$\begin{aligned} Y_m = {} & A_m\{e^{m\zeta} \ln 2m\zeta - e^{-m\zeta}\,\mathrm{Ei}(2m\zeta)\} \\ & + B_m\{e^{-m\zeta} \ln 2m\zeta - e^{m\zeta}\,\mathrm{Ei}(-2m\zeta)\} + C_m e^{m\zeta} + D_m e^{-m\zeta}. \end{aligned} \tag{5.57}$$

The constants A_m, B_m, C_m, D_m may now be determined from the boundary conditions along the edges $y = y_0, y_0 + b$.

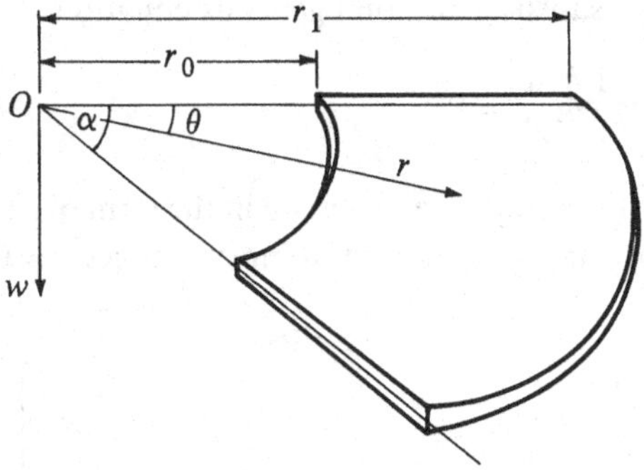

Fig. 5.3

5.5 Circular plates

The governing differential equation in polar coordinates for plates with variable rigidity may be obtained by substituting (3.1) and (3.2) into (1.28b) to yield

$$\nabla^2(D\nabla^2 w) - (1-\nu)\left\{\frac{\partial^2 D}{\partial r^2}\left(\frac{1}{r}\frac{\partial w}{\partial r} + \frac{1}{r^2}\frac{\partial^2 w}{\partial \theta^2}\right)\right.$$
$$\left. - 2\frac{\partial}{\partial r}\left(\frac{1}{r}\frac{\partial D}{\partial \theta}\right)\cdot\frac{\partial}{\partial r}\left(\frac{1}{r}\frac{\partial w}{\partial \theta}\right) + \frac{\partial^2 w}{\partial r^2}\left(\frac{1}{r}\frac{\partial D}{\partial r} + \frac{1}{r^2}\frac{\partial^2 D}{\partial \theta^2}\right)\right\} = q. \tag{5.58}$$

When there is rotational symmetry in D, q and in the boundary conditions, it is possible to obtain a wide range of solutions to (5.58). Without such symmetry, known solutions are restricted to variations of D proportional to r^k (see Mansfield 1962). Solutions for $k = 1, 2, 3$ are given below for a plate in the form of a sector bounded by the lines $\theta = 0, \alpha$ and $r = r_0, r_1$.

5.5.1 *Sector plate with rigidity varying as r* (see Fig. 5.3)

If the rigidity varies as

$$D = D_1\frac{r}{r_1}, \tag{5.59}$$

equation (5.58) becomes

$$\nabla^2(r\nabla^2 w) - (1-\nu)\frac{1}{r}\frac{\partial^2 w}{\partial r^2} = q\frac{r_1}{D_1}. \tag{5.60}$$

In searching for a solution of (5.60), we take $w = w_1 + w_2$, where w_1 is a particular integral and w_2 satisfies the homogeneous equation

$$\nabla^2(r\nabla^2 w_2) - (1-\nu)\frac{1}{r}\frac{\partial^2 w_2}{\partial r^2} = 0. \tag{5.61}$$

The general solution of (5.61) appropriate to a plate in the form of a sector subtending an angle α and simply supported along the edges $\theta = 0, \alpha$ is given by

$$\left.\begin{aligned} w_2 &= \sum_{n=1}^{\infty} R_n(\varrho)\sin\frac{n\pi\theta}{\alpha}, \\ &\text{where } \varrho = r/r_1, \\ R_n(\varrho) &= \sum_{i=1}^{4} A_{n,i}\varrho^{m_i} \\ &\text{and the } m_i \text{ are the roots of the equation} \\ &(m^2 - n^2\pi^2/\alpha^2)\{(m-1)^2 - n^2\pi^2/\alpha^2\} - m(m-1)(1-\nu) = 0. \end{aligned}\right\} \tag{5.62}$$

When the applied loading can be expressed in the form

$$q = \varrho^{\lambda}\sum_{n=1}^{\infty} q_n \sin\frac{n\pi\theta}{\alpha}, \tag{5.63}$$

it can be shown by substitution into (5.60) that the function w_1 is given by

$$\left.\begin{aligned} w_1 &= \frac{r_1^4}{D_1}\rho^{\lambda+3}\sum_{n=1}^{\infty}\frac{q_n}{K_n}\sin\frac{n\pi\theta}{\alpha}, \\ &\text{where} \\ K_n &= \{(\lambda+3)^2 - n^2\pi^2/\alpha^2\}\{(\lambda+2)^2 - n^2\pi^2/\alpha^2\} \\ &\quad - (1-\nu)(\lambda+3)(\lambda+2). \end{aligned}\right\} \tag{5.64}$$

Expression (5.64) above satisfies the boundary conditions for a plate simply supported along the sides $\theta = 0, \alpha$. The coefficients $A_{n,1}, A_{n,2}, A_{n,3}, A_{n,4}$ can now be determined from the boundary conditions along $r = r_0, r_1$.

5.5.2 *Sector plate with rigidity varying as r^2*

This case was also discussed by Gran Olsson (1939). If the rigidity varies as

$$D = D_1\left(\frac{r}{r_1}\right)^2, \tag{5.65}$$

equation (5.58) assumes the simple form

$$\nabla^2\{r^2\nabla^2 w - 2(1-\nu)w\} = qr_1^2/D_1. \tag{5.66}$$

In searching for a solution of (5.66), we again take $w = w_1 + w_2$, where w_1 is a particular integral and w_2 satisfies the homogeneous equation

$$\nabla^2\{r^2\nabla^2 w_2 - 2(1-\nu)w_2\} = 0. \tag{5.67}$$

The general solution of (5.67) appropriate to a plate simply supported along the edges $\theta = 0, \alpha$ is given by

$$w_2 = \sum_{n=1}^{\infty} R_n(\varrho)\sin\frac{n\pi\theta}{\alpha},$$

where $\varrho = r/r_1$, and

$$R_n(\varrho) = A_n\varrho^{n\pi/\alpha} + B_n\varrho^{-n\pi/\alpha} + C_n\varrho^{m\pi/\alpha} + D_n\varrho^{-m\pi/\alpha}$$

and

$$m^2 = \{n^2 + 2(1-\nu)\alpha^2/\pi^2\}. \tag{5.68}$$

If the applied loading can be expressed in the form of (5.63), it can be shown that the function w_1 which satisfies the conditions of simple support along the edges $\theta = 0, \alpha$ is given by

$$w_1 = \frac{r_1^4}{D_1}\varrho^{\lambda+2}\sum_{n=1}^{\infty}\frac{q_n}{K_n}\sin\frac{n\pi\theta}{\alpha},$$

where

$$K_n = \{(\lambda+2)^2 - n^2\pi^2/\alpha^2\}\{(\lambda+2)^2 - 2(1-\nu) - n^2\pi^2/\alpha^2\}. \tag{5.69}$$

5.5.3 *Sector plate with rigidity varying as r^3*

This case is of particular interest as it corresponds to the thickness varying directly as r. If

$$D = D_1\left(\frac{r}{r_1}\right)^3, \tag{5.70}$$

equation (5.58) assumes the form

$$\nabla^2(r\nabla^2 w) - 3(1-\nu)\left(r\frac{\partial^2 w}{\partial r^2} + 2\frac{\partial w}{\partial r} + \frac{2}{r}\frac{\partial^2 w}{\partial\theta^2}\right) = qr_1^3/D_1. \tag{5.71}$$

Following a similar analysis to that of Sections 5.5.1 and 5.5.2, we may express the solution of (5.71) in the form

$$w_2 = \sum_{n=1}^{\infty} R_n(\varrho)\sin\frac{n\pi\theta}{\alpha},$$

where

$$\varrho = r/r_1,$$

$$R_n(\varrho) = \sum_{i=1}^{4} A_{n,i}\varrho^{m_i} \tag{5.72}$$

and the m_i are roots of the equation

$$\left.\begin{aligned} m^4 + 2m^3 - m^2(2 - 3\nu + 2n^2\pi^2/\alpha^2) \\ - m(3 - 3\nu + 2n^2\pi^2/\alpha^2) \\ + (5 - 6\nu)n^2\pi^2/\alpha^2 + n^4\pi^4/\alpha^4 = 0. \end{aligned}\right\}$$

Further, if the applied loading is of the form (5.63), we find

$$\left.\begin{aligned} & w_1 = \frac{r_1^4}{D_1}\varrho^{\lambda+1}\sum_{n=1}^{\infty}\frac{q_n}{K_n}\sin\frac{n\pi\theta}{\alpha}, \\ & \text{where} \\ & K_n = (\lambda+1)^4 + 2(\lambda+1)^3 - (\lambda+1)^2(2 - 3\nu + 2n^2\pi^2/\alpha^2) \\ & \quad - (\lambda+1)(3 - 3\nu + 2n^2\pi^2/\alpha^2) \\ & \quad + (5 - 6\nu)n^2\pi^2/\alpha^2 + n^4\pi^4/\alpha^4. \end{aligned}\right\} \tag{5.73}$$

5.6 Circular plates with rotational symmetry

When there is rotational symmetry in D, q and in the boundary conditions, the deflexion is likewise independent of θ and (5.58) becomes

$$\left.\begin{aligned} & \frac{1}{r}\frac{\mathrm{d}}{\mathrm{d}r}\left[r\frac{\mathrm{d}}{\mathrm{d}r}\left\{D\left(\frac{\mathrm{d}\varphi}{\mathrm{d}r} + \frac{\varphi}{r}\right)\right\}\right] - \frac{(1-\nu)}{r}\frac{\mathrm{d}}{\mathrm{d}r}\left(\varphi\frac{\mathrm{d}D}{\mathrm{d}r}\right) = -q, \\ & \text{where} \\ & \varphi = -\frac{\mathrm{d}w}{\mathrm{d}r}. \end{aligned}\right\} \tag{5.74}$$

Equation (5.74) may be multiplied by r and integrated once to give

$$\left.\begin{aligned} & D\frac{\mathrm{d}}{\mathrm{d}\varrho}\left(\frac{\mathrm{d}\varphi}{\mathrm{d}\varrho} + \frac{\varphi}{\varrho}\right) + \frac{\mathrm{d}D}{\mathrm{d}\varrho}\left(\frac{\mathrm{d}\varphi}{\mathrm{d}\varrho} + \nu\frac{\varphi}{\varrho}\right) = -\frac{r_1^3}{\varrho}\int_0^{\varrho} q\varrho\,\mathrm{d}\varrho, \\ & \text{where} \\ & \varrho = r/r_1. \end{aligned}\right\} \tag{5.75}$$

Variations of D for which it is possible to obtain closed form solutions of (5.75) have been summarized by Conway (1953) and are listed below. In most cases only the complementary solution of (5.75) is given, for the particular integral may then be obtained by the method of *variation of parameters*, as discussed, for example, in Jeffreys and Jeffreys (1950).

5.6.1 *Rigidity varying as ϱ^k*

The complementary solution of (5.75) satisfies the equation

$$\varrho^2\frac{\mathrm{d}^2\varphi}{\mathrm{d}\varrho^2} + (k+1)\varrho\frac{\mathrm{d}\varphi}{\mathrm{d}\varrho} - (1 - \nu k)\varphi = 0 \tag{5.76}$$

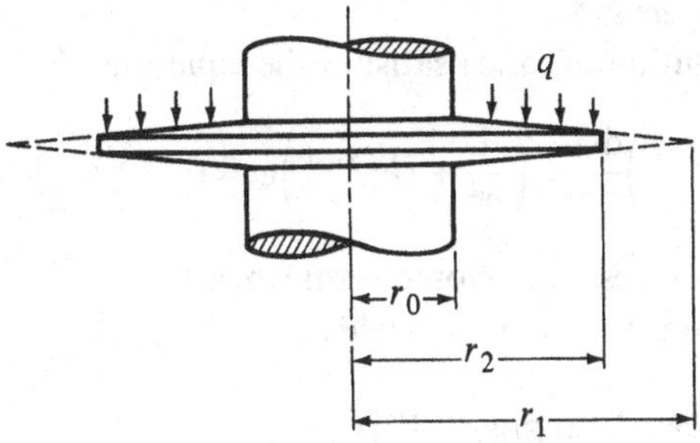

Fig. 5.4

whose solution is

$$\left.\begin{aligned}\varphi &= A\varrho^{-\lambda-\frac{1}{2}k} + B\varrho^{\lambda-\frac{1}{2}k},\\ &\text{where}\\ \lambda^2 &= 1 - \nu k + \tfrac{1}{4}k^2.\end{aligned}\right\} \tag{5.77}$$

5.6.2 *Rigidity varying as* $(1-\varrho^k)^n$

If we write

$$\beta = \varrho^k,$$

the complementary solution satisfies the equation

$$\beta(1-\beta)\frac{d^2\varphi}{d\beta^2} + (1-\beta-n\beta)\frac{d\varphi}{d\beta} - \left(\frac{1-\beta}{k^2\beta} + \frac{\nu n}{k}\right)\varphi = 0 \tag{5.78}$$

whose general solution can be expressed in terms of hypergeometric functions. A case of particular interest occurs when $k = 1$, $n = 3$ which corresponds to a plate with linearly varying thickness (Fig. 5.4). With these values for k and n and taking $\nu = 1/3$, it is possible to integrate (5.78) to give

$$\varphi = A\frac{1+2\varrho}{\varrho} + B\frac{3\varrho - 2\varrho^2}{(1-\varrho)^2}. \tag{5.79}$$

The particular integral φ' has been determined by Conway (1951) for a uniform loading q. With the notation shown in Fig. 5.4,

$$\varphi' = \frac{qr_1^3}{12D_0}\left[\frac{(4\varrho^3 + 15\varrho^2 - 6\varrho - 6)}{6\varrho(1-\varrho)^2} - \frac{\varrho_2^2(2\varrho^2 + \varrho - 1)}{\varrho(1-\varrho)^2} - \frac{(2\varrho+1)(1+\varrho_2^2)}{\varrho}\ln(1-\varrho) - \frac{(3-2\varrho)\varrho\varrho_2^2}{(1-\varrho)^2}\ln\varrho\right]. \tag{5.80}$$

The constants A, B may now be determined from the boundary conditions.

5.6.3 *Rigidity varying as $e^{-\varrho^k}$*

The complementary solution of (5.75) satisfies the equation

$$\frac{d^2\varphi}{d\varrho^2}+\left(\frac{1}{\varrho}-k\varrho^{k-1}\right)\frac{d\varphi}{d\varrho}-\left(\frac{1}{\varrho^2}+\nu k\varrho^{k-2}\right)\varphi=0 \tag{5.81}$$

whose solution may be expressed in closed form when $1/\nu = 1, 3, 5, \ldots$ and $k = 2\nu$. Taking $\nu = \frac{1}{3}, k = \frac{2}{3}$, for example, yields

$$\varphi=\frac{1}{\varrho}\{A(2-\varrho^{2/3})e^{\varrho^{2/3}}+B(2+\varrho^{2/3})\}. \tag{5.82}$$

References

Conway, H. D. Axially symmetrical plates with linearly varying thickness. *J. Appl. Mech.*, **18**, pp. 140–2 (June 1951).

______. Closed form solutions for plates of variable thickness. *J. Appl. Mech.*, **20**, p. 564 (December 1953).

______. A Levy-type solution for a rectangular plate of variable thickness. *J. Appl. Mech.*, **25**, pp. 297–8 (June 1958).

Duncan, W. J. The torsion and flexure of cylinders and tubes. *Aero. Res. Council R. & M.* No. 1444. H.M.S.O. (February 1932).

Gran Olsson, R. *Ingenieur-Archiv*, **5**, p. 363 (1934); **10** (1939).

Jeffreys, H., and Jeffreys, B. S. *Methods of mathematical physics*. 2nd ed. Cambridge, 1950.

Mansfield, E. H. On the analysis of elastic plates of variable thickness. *Quart. J. Mech. and Appl. Math.*, **15**, pp. 167–92 (1962).

Additional references

Mansfield, E. H. The influence of aerodynamic heating on the flexural rigidity of a thin wing. *Aero. Res. Council R. & M.* No. 3115. H.M.S.O. (September 1957).

______. On the deflexion of an anisotropic cantilever plate with variable rigidity. *Proc. R. Soc. Lond.* Ser. A., **366**, pp. 491–515 (1979).

Fung, Y. C. Bending of thin elastic plates of variable thickness. *J. Aero. Sci.*, **20**, p. 455 (1953).

6

Approximate methods

There are many problems concerning the small-deflexion bending of plates for which a rigorous solution is impracticable, and recourse must then be had to approximate methods of analysis. Broadly speaking, such methods fall into three categories: first, those that start from the governing differential equation whose approximate solution is obtained by numerical integration; second, those based on a finite element analysis (see, for example, Zienciewicz 1977); and third, continuum solutions that are based on the principle of minimum potential energy, or on allied energy principles. In this chapter we consider methods in this third category, and as a preliminary measure we determine the strain energy of a deformed plate.

6.1 The strain energy of an isotropic plate

The strain energy of a deformed isotropic plate may be regarded as the sum of that due to bending and that due to stretching of the middle surface. That due to bending will now be determined.

6.1.1 The strain energy of bending

The strain energy of bending per unit area of a deformed place, U_b', is expressed most simply in terms of the principal moments M_1, M_2 and the principal curvatures κ_1, κ_2. Thus,

$$\left.\begin{aligned} U_b' &= \tfrac{1}{2}(M_1\kappa_1 + M_2\kappa_2) \\ &= \tfrac{1}{2}D(\kappa_1^2 + 2\nu\kappa_1\kappa_2 + \kappa_2^2) \end{aligned}\right\} \tag{6.1}$$

in virtue of (1.9).

The total strain energy of bending, U_b is obtained by integrating U_b' over the entire plate. It is possible to include the effect of a variable rigidity by keeping D under the integral sign. Thus

$$U_b = \frac{1}{2}\iint D(\kappa_1^2 + 2\nu\kappa_1\kappa_2 + \kappa_2^2)\mathrm{d}x\mathrm{d}y. \tag{6.2}$$

Now there are certain advantages in expressing (6.2) in each of two alternative forms. The first form is obtained by writing

$$U_b = \frac{1}{2}\iint \frac{t^3}{12}\left\{\frac{E}{2(1-\nu)}(\kappa_1+\kappa_2)^2 + \frac{E}{2(1+\nu)}(\kappa_1-\kappa_2)^2\right\}\mathrm{d}x\,\mathrm{d}y, \tag{6.2a}$$

where each term in the braces is necessarily positive. The first term represents the strain energy due to each element of the plate undergoing a spherical curvature equal to its average curvature; the second term represents the strain energy due to each element undergoing its maximum twisting curvature. Expression (6.2a) can be used to obtain upper and lower bounds for the effect of the Poisson ratio on the total strain energy of bending of a deformed plate – a problem of some interest in the field of buckling and vibrations.

If two similar plates with equal values of E, t but differing values of ν undergo the same deformation, their bending strain energies U_{b1}, U_{b2} satisfy the following inequality:

$$\frac{1-\nu_1}{1-\nu_2} < \frac{U_{b1}}{U_{b2}} < \frac{1+\nu_1}{1+\nu_2} \tag{6.3}$$

in which it is tacitly assumed that $\nu_1 > \nu_2$.

The second form for (6.2) is given by

$$U_b = \frac{1}{2}\iint D\{(\kappa_1+\kappa_2)^2 - 2(1-\nu)\kappa_1\kappa_2\}\,\mathrm{d}x\,\mathrm{d}y \tag{6.2b}$$

which can be readily expressed in terms of the second derivatives of w in virtue of (1.14) and (1.16):

$$\begin{aligned} U_b &= \frac{1}{2}\iint D\left[\left(\frac{\partial^2 w}{\partial x^2}+\frac{\partial^2 w}{\partial y^2}\right)^2 \right.\\ &\qquad \left. -2(1-\nu)\left\{\frac{\partial^2 w}{\partial x^2}\frac{\partial^2 w}{\partial^2 y} - \left(\frac{\partial^2 w}{\partial x \partial y}\right)^2\right\}\right]\mathrm{d}x\,\mathrm{d}y \\ &= \frac{1}{2}\iint D\{(\nabla^2 w)^2 - (1-\nu)\Diamond^4(w,w)\}\,\mathrm{d}x\,\mathrm{d}y. \end{aligned} \tag{6.4}$$

Simple expression for U_b if D is constant or linearly varying. It was shown in Sections 1.4.1 and 5.4 that if D is constant or linearly varying, the governing differential equation of the deflexion is independent of ν. For such plates ν can only affect the deflexion, and hence U_b, through its influence on the boundary conditions; if these conditions are purely kinematic or are otherwise independent of ν, we can then deduce that

$$\iint D\diamondsuit^4(w, w)\,dx\,dy = 0 \tag{6.5}$$

so that

$$U_b = \frac{1}{2}\iint D(\nabla^2 w)^2\,dx\,dy. \tag{6.6}$$

From the results of Section 1.7 it will be seen that for curved boundaries (6.6) is valid only when they are clamped, but if the boundaries are straight, (6.6) is valid whether they are clamped, simply supported or elastically restrained against rotation.

Referring back to the earlier discussion on the effect of the Poisson ratio on the strain energy of bending, it will be seen that while (6.3) is generally true for plates with equal values of E, t, there are many instances in which the influence of ν on U_b may be precisely defined. Thus, whenever (6.6) is valid,

$$\frac{U_{b2}}{U_{b1}} = \frac{1-\nu_1^2}{1-\nu_2^2}. \tag{6.7}$$

6.1.2 *Strain energy due to middle-surface forces*

When middle-surface forces N_x, N_y, N_{xy} are present, the initial strain energy due to these middle-surface forces, $U_{\Phi,0}$, will change when the plate deflects. To determine this change in strain energy it is convenient to denote by u, v, w the displacements relative to the plate subjected to middle-surface forces alone. It can be shown from geometrical considerations that the changes in the middle-surface strains are now given by

$$\left.\begin{aligned}
\delta\varepsilon_x &= \frac{\partial u}{\partial x} + \frac{1}{2}\left(\frac{\partial w}{\partial x}\right)^2, \\
\delta\varepsilon_y &= \frac{\partial v}{\partial y} + \frac{1}{2}\left(\frac{\partial w}{\partial y}\right)^2, \\
\delta\varepsilon_{xy} &= \frac{\partial u}{\partial y} + \frac{\partial v}{\partial x} + \frac{\partial w}{\partial x}\frac{\partial w}{\partial y}.
\end{aligned}\right\} \tag{6.8}$$

Now we are assuming that the deflexions are sufficiently small for the middle-surface forces to remain sensibly constant, so that the change in strain energy δU_Φ is given by

$$\delta U_\Phi = \iint (N_x\delta\varepsilon_x + N_y\delta\varepsilon_y + N_{xy}\delta\varepsilon_{xy})\,dx\,dy$$

$$= \iint \left\{ N_x \frac{\partial u}{\partial x} + N_y \frac{\partial v}{\partial y} + N_{xy}\left(\frac{\partial u}{\partial y} + \frac{\partial v}{\partial x} \right) \right\} \mathrm{d}x\,\mathrm{d}y$$
$$+ \frac{1}{2} \iint \left\{ N_x \left(\frac{\partial w}{\partial x} \right)^2 + N_y \left(\frac{\partial w}{\partial y} \right)^2 + 2N_{xy} \frac{\partial w}{\partial x}\frac{\partial w}{\partial y} \right\} \mathrm{d}x\,\mathrm{d}y \tag{6.9}$$

by virtue of (6.8).

The first term in (6.9) may be equated with the work done by the middle-surface forces acting around the boundary, W_Φ. The potential energy of the middle-surface forces Π_Φ is thus given – apart from the constant term $U_{\Phi,0}$ – by

$$\Pi_\Phi = \delta U_\Phi - W_\Phi$$
$$= \frac{1}{2} \iint \left\{ N_x \left(\frac{\partial w}{\partial x} \right)^2 + N_y \left(\frac{\partial w}{\partial y} \right)^2 + N_{xy} \frac{\partial w}{\partial x}\frac{\partial w}{\partial y} \right\} \mathrm{d}x\,\mathrm{d}y$$
$$= \frac{1}{2} \int \{ \tfrac{1}{2} \lozenge^4(w^2, \Phi) - w \lozenge^4(w, \Phi) \} \, \mathrm{d}x\,\mathrm{d}y \tag{6.10}$$

in virtue of (1.33) and the definition of the operator $\lozenge^4$.

The potential energy of the distributed loading $q(x, y)$ is given by

$$\Pi_q = - \iint q w \, \mathrm{d}x\,\mathrm{d}y \tag{6.11}$$

and the potential energy of externally applied moments and shears is given by

$$\Pi_e = \oint \left(Q_n w - M_n \frac{\partial w}{\partial \boldsymbol{n}} - M_{ns} \frac{\partial w}{\partial s} \right) \mathrm{d}s, \tag{6.12}$$

where $\boldsymbol{n}$ is the outward normal to the boundary and s is directed along the boundary.

Similarly, it may be shown that if the edges of the plate are elastically restrained against rotation, the strain energy U_e stored in the surrounding structure is given by

$$U_e = \frac{1}{2} \oint \chi \left(\frac{\partial w}{\partial \boldsymbol{n}} \right)^2 \mathrm{d}s \tag{6.13}$$

and if the boundary of the plate is elastically restrained against deflexion

$$U_e = \frac{1}{2} \oint \varrho w^2 \, \mathrm{d}s, \tag{6.13a}$$

where χ and ϱ are defined in Section 1.7.

The total potential energy Π is the sum

$$\Pi = U_b + \Pi_\Phi + \Pi_q + \Pi_e + U_e. \tag{6.14}$$

6.2 Strain energy in multi-layered anisotropic plate

In the general case in which there is coupling between moments and planar strains, the total strain energy per unit area U' due to bending and middle-surface forces may be obtained by integrating the strain energy density through the thickness. Thus from Section 1.8,

$$\begin{aligned} U' &= \frac{1}{2}\int_{-\frac{1}{2}t}^{\frac{1}{2}t} \boldsymbol{\sigma}^{\mathrm{T}} \boldsymbol{\varepsilon}\, \mathrm{d}z, \\ &= \frac{1}{2}\int_{-\frac{1}{2}t}^{\frac{1}{2}t} \boldsymbol{\sigma}^{\mathrm{T}} (\boldsymbol{\varepsilon}^0 + z\boldsymbol{\kappa})\, \mathrm{d}z, \\ &= \tfrac{1}{2}(\mathbf{N}^{\mathrm{T}} \boldsymbol{\varepsilon}^0 + \mathbf{M}^{\mathrm{T}} \boldsymbol{\kappa}). \end{aligned} \tag{6.15}$$

Equation (6.15) could have been written down directly; the element of coupling appears when we express **N** and **M** in terms of middle-surface strains and plate curvatures, for example. Thus, combining (6.15) with (1.94) yields

$$U' = \tfrac{1}{2}(\boldsymbol{\varepsilon}^{0,\mathrm{T}} \mathbf{A} \boldsymbol{\varepsilon}^0 + 2\boldsymbol{\kappa}^{\mathrm{T}} \mathbf{B} \boldsymbol{\varepsilon}^0 + \boldsymbol{\kappa}^{\mathrm{T}} \mathbf{D} \boldsymbol{\kappa}). \tag{6.16}$$

Similarly, if we express U' in terms of **N** and $\boldsymbol{\kappa}$ we find, after some simplification,

$$U' = \tfrac{1}{2}(\mathbf{N}^{\mathrm{T}} \mathbf{A}^{-1} \mathbf{N} + \boldsymbol{\kappa}^{\mathrm{T}}\, \mathbf{d} \boldsymbol{\kappa}), \tag{6.17}$$

where **d**, defined by (1.96), embodies the element of coupling.

6.2.1 *Zero coupling between* N *and* M

When **B** is zero, there is no coupling between moments and planar strains, and the strain energy per unit area is given simply by

$$\begin{aligned} U' &= U_\Phi + U_b, \\ &= \tfrac{1}{2}(\boldsymbol{\varepsilon}^{0,\mathrm{T}} \mathbf{A} \boldsymbol{\varepsilon}^0 + \boldsymbol{\kappa}^{\mathrm{T}} \mathbf{D} \boldsymbol{\kappa}). \end{aligned} \tag{6.18}$$

From (6.15) the strain energy due to the middle-surface forces may also be expressed in the form

$$U_\Phi = \frac{1}{2}\iint \mathbf{N}^{\mathrm{T}} \boldsymbol{\varepsilon}^0\, \mathrm{d}x\, \mathrm{d}y, \tag{6.19}$$

which is the same as that for the isotropic plate, and it follows that the analysis of Section 6.1.2 is also valid for the uncoupled anisotropic plate.

From (6.18) and (1.94) the strain energy of bending is given by

$$U_b = \frac{1}{2}\iint \boldsymbol{\kappa}^{\mathrm{T}}\mathbf{D}\boldsymbol{\kappa}\,\mathrm{d}x\,\mathrm{d}y,$$

$$= D_{11}\left(\frac{\partial^2 w}{\partial x^2}\right)^2 + 2D_{12}\frac{\partial^2 w}{\partial x^2}\frac{\partial^2 w}{\partial y^2} + D_{22}\left(\frac{\partial^2 w}{\partial y^2}\right)^2$$

$$+ 4\frac{\partial^2 w}{\partial x \partial y}\left(D_{16}\frac{\partial^2 w}{\partial x^2} + D_{26}\frac{\partial^2 w}{\partial y^2} + D_{66}\frac{\partial^2 w}{\partial x \partial y}\right). \tag{6.20}$$

6.3 Principle of minimum total potential energy – Ritz method

The principle of minimum total potential energy may now be applied to obtain an approximate solution to plate problems. In this application it is sometimes known as the Ritz method. A form for the deflexion is chosen which satisfies the boundary conditions and which contains a number of disposable parameters. Thus, we may take a linear combination of the form

$$w = \sum_{n=1}^{N} B_n w_n(x, y) \tag{6.21}$$

or, more generally,

$$w = \sum_{m=1}^{M}\sum_{n=1}^{N} B_{mn} w_{mn}(x, y) \tag{6.22}$$

where the parameters B_{mn} are determined from the MN equations

$$\frac{\partial \Pi}{\partial B_{mn}} = 0. \tag{6.23}$$

When the series of functions w_{mn} in (6.22) is sufficiently general to represent all possible displacement patterns, the solution will tend to the correct one as M, N increase. Thus, for the clamped rectangular plate with constant rigidity Ritz (1911) chose the doubly infinite set

$$w_{mn} = F_m(x)F_n(y),$$

where the functions F are the modes of a clamped beam. Pickett (1939) chose the doubly infinite set

$$w_{mn} = (a^2 - 4x^2)^2(b^2 - 4y^2)^2 x^m y^n, \tag{6.24}$$

where the origin is at the centre of the plate, and he has shown how the corresponding parameters B_{mn} may be determined for an arbitrary load distribution. These expressions are not the only ones that may be used for the investigation of a clamped rectangular plate and the following are also suitable

$$w_{mn} = \sin\frac{\pi x}{a}\sin\frac{\pi y}{b}\sin\frac{m\pi x}{a}\sin\frac{n\pi y}{b} \tag{6.25}$$

or

$$w_{mn} = X_m Y_n,$$

where

$$\left.\begin{aligned} X_m &= \frac{x}{a}\left(\frac{x}{a}-1\right)^2 + (-1)^m\frac{x^2}{a^2}\left(\frac{x}{a}-1\right) - \frac{1}{m\pi}\sin\frac{m\pi x}{a} \\ Y_n &= \frac{y}{b}\left(\frac{y}{b}-1\right)^2 + (-1)^n\frac{y^2}{b^2}\left(\frac{y}{b}-1\right) - \frac{1}{n\pi}\sin\frac{n\pi y}{b}. \end{aligned}\right\} \tag{6.26}$$

The functions X_m and Y_n in (6.26) are known as Iguchi functions (1938), and they have been generalized by Hopkins (1945, p. 51) to cover the case when opposite edges are clamped and simply supported. The appropriate functions X_m, Y_n are then obtained from (2.3) by suitable choice of the coefficients a_m, b_m, and so on. Thus, if the edge at $x = 0$ is clamped and the edge at $x = a$ is simply supported:

$$X_m = \frac{x}{a}\left(\frac{x}{a}-1\right)\left(\frac{x}{2a}-1\right) - \frac{1}{m\pi}\sin\frac{m\pi x}{a}, \tag{6.27}$$

while if the edge at $x = 0$ is simply supported and that at $x = a$ is clamped:

$$X_m = \frac{x}{2a}\left(\frac{x^2}{a^2}-1\right)(-1)^m - \frac{1}{m\pi}\sin\frac{m\pi x}{a} \tag{6.28}$$

and there are analogous expressions for Y_n. When opposite edges are simply supported, the sine terms alone are sufficient to satisfy the boundary conditions and it is generally possible to obtain a solution by 'exact' methods. If the edges at $x = 0, a$ are elastically restrained against rotation, so that the boundary condition is given by (1.72), the functions X_m are given by

$$X_m = b_m\left(\frac{x}{a}\right) + c_m\left(\frac{x}{a}\right)^2 + d_m\left(\frac{x}{a}\right)^3 - \frac{1}{m\pi}\sin\frac{m\pi x}{a},$$

where

$$\left.\begin{aligned} b_m &= \frac{4\lambda_0 + \lambda_0\lambda_a - 2\lambda_a(-1)^m}{(4+\lambda_0)(4+\lambda_a)-4}, \\ c_m &= -\lambda_0\left(\frac{6+\lambda_a\{2+(-1)^m\}}{(4+\lambda_0)(4+\lambda_a)-4}\right), \\ d_m &= \frac{2\{\lambda_0 + \lambda_a(-1)^m\} + \lambda_0\lambda_a\{1+(-1)^m\}}{(4+\lambda_0)(4+\lambda_a)-4} \end{aligned}\right\} \tag{6.29}$$

and

$$\left.\lambda_0 = a\chi_0/D, \qquad \lambda_a = a\chi_a/D.\right|$$

There is an analogous expression for Y_n with x/a replaced by y/b, and so on.

Finally, we note that for plates whose entire boundaries are clamped, further methods of analysis are available, as discussed in Sections 6.6 and 6.7. Examples have already been presented in Sections 2.3.1 and 3.7.

6.3.1 *Application of the energy method in buckling problems*

The magnitude of middle-surface forces necessary to cause buckling may also be estimated by considerations of energy. During buckling the work done by the middle-surface forces W_Φ is equal to the increase in strain energy of the plate $(U + \delta U_\Phi)$ plus that in the surrounding structure U_e. Thus, from (6.4), (6.9) and (6.10) we find

$$-\tfrac{1}{2}\iint\left\{N_x\left(\frac{\partial w}{\partial x}\right)^2 + N_y\left(\frac{\partial w}{\partial y}\right)^2 + 2N_{xy}\frac{\partial w}{\partial x}\frac{\partial w}{\partial y}\right\}\mathrm{d}x\,\mathrm{d}y$$
$$= \tfrac{1}{2}\iint D\{(\nabla^2 w)^2 - (1-\nu)\Diamond^4(w,w)\}\,\mathrm{d}x\,\mathrm{d}y + U_e. \tag{6.30}$$

If the boundaries are simply supported, clamped or free, no strain energy is stored by the surrounding structure and $U_e = 0$.

In using (6.30) to determine the magnitude of the middle-surface forces necessary to cause buckling, it is convenient to write

$$N_x = -\gamma N'_x, \quad N_y = -\gamma N'_y, \quad N_{xy} = -\gamma N'_{xy}$$

so that positive values of γ, N'_x, N'_y correspond to compressive middle-surface forces. Further, by varying γ the middle-surface forces are varied in proportion to their magnitudes. The onset of buckling is to be determined from the condition that γ is a minimum; that is, we must minimize the expression

$$\gamma = \frac{\displaystyle\iint D\{(\nabla^2 w)^2 - (1-\nu)\Diamond^4(w,w)\}\,\mathrm{d}x\,\mathrm{d}y + 2U_e}{\displaystyle\iint\left\{N'_x\left(\frac{\partial w}{\partial x}\right)^2 + N'_y\left(\frac{\partial w}{\partial y}\right)^2 + 2N'_{xy}\frac{\partial w}{\partial x}\frac{\partial w}{\partial y}\right\}\mathrm{d}x\,\mathrm{d}y}. \tag{6.31}$$

Buckling of plate with variable rigidity. As an example in the use of (6.31) the buckling of a simply supported square plate under uniform compressive forces N_x is considered. The rigidity of the plate varies linearly from D_0

at $x = 0$ to D_1 at $x = a$. The deflected form is represented by the single series

$$w = \sin\frac{\pi y}{a}\sum_{m=1}^{\infty} c_m \sin\frac{m\pi x}{a}. \tag{6.32}$$

In the choice of a single sinusoidal variation with y we are guided initially by experience which can be confirmed by noting that a deflexion of the form $f(x)\sin(\pi y/a)$ can be made to satisfy the governing differential equation. Thus, expression (6.32) tends to the correct buckled form as m tends to infinity. The choice of sinusoidal terms ensures that the boundary conditions are satisfied.

Substitution of (6.32) in (6.31) and integration yields

$$-\frac{a^2}{\pi^2}N_x = \frac{\frac{1}{2}(D_0 + D_1)\sum_{m=1}^{\infty}(1+m^2)^2 c_m^2 + 2(D_1 - D_0)I}{\sum_{m=1}^{\infty} m^2 c_m^2}, \tag{6.33}$$

where

$$I = \frac{1}{\pi^2}\sum_{m=1}^{\infty}\sum_{\substack{r=1\\ r\neq m}}^{\infty} c_m c_r (1+m^2)(1+r^2) \times\left(\frac{1-(-1)^{m+r}}{(m+r)^2} - \frac{1-(-1)^{m+r}}{(m-r)^2}\right).$$

The condition that (6.33) is to be minimized with respect to the coefficients c_m results in an infinite system of simultaneous linear equations,

$$-\frac{a^2}{\pi^2}N_x m^2 c_m = \frac{1}{2}(D_0 + D_1)(1+m^2)^2 c_m - \frac{4}{\pi^2}(D_1 - D_0)\sum_{\substack{r=1\\ r\neq m}}^{\infty}\frac{mr(1+m^2)(1+r^2)\{1-(-1)^{m+r}\}c_r}{(m^2-r^2)^2}. \tag{6.34}$$

If only one term in the series is taken, we find, not surprisingly,

$$-N_x = \frac{2\pi^2(D_0 + D_1)}{a^2},$$

but if two terms are taken,

$$\left(\frac{a^2 N_x}{\pi^2(D_0 + D_1)} + 2\right)\left(\frac{a^2 N_x}{\pi^2(D_0 + D_1)} + \frac{25}{8}\right) = \frac{6400}{81\pi^4}\left(\frac{D_0 - D_1}{D_0 + D_1}\right)^2 \tag{6.35}$$

which may be solved to give

$$-N_x = \frac{\pi^2(D_0+D_1)}{16a^2}\left[41 - 9\left\{1 + 2.57\left(\frac{D_0-D_1}{D_0+D_1}\right)^2\right\}^{\frac{1}{2}}\right]. \tag{6.36}$$

6.4 The Galerkin method

In the Galerkin method (see, for example, Duncan 1937) the deflexion is again represented by (6.22) and the functions $w_{mn}(x, y)$ are chosen to satisfy the boundary conditions, but the parameters β_{mn} are determined from the following system of equations, thus obviating the necessity to determine the potential energy:

$$\iint \{D\nabla^4 w - \Diamond^4(\Phi, w) - q\} w_{mn}\, dx\, dy = 0, \quad \begin{pmatrix} m = 1, 2, \ldots, M \\ n = 1, 2, \ldots, N \end{pmatrix}. \tag{6.37}$$

If the rigidity of the plate varies (6.37) is replaced by the equation

$$\iint \{\nabla^2(D\nabla^2 w) - (1-\nu)\Diamond^4(D, w) - \Diamond^4(\Phi, w) - q\} w_{mn}\, dx\, dy = 0. \tag{6.38}$$

As a possible application of these equations, the deflexion of a rectangular plate whose boundaries are elastically restrained against rotation might be considered, using the functions X_m, Y_n of the previous section. As a simpler example the effect of shearing forces N_{xy} on the deflexion of a simply supported square plate will be considered here.

6.4.1 *Effect of shearing forces N_{xy} on plate deflexion*

Substituting (2.1) and (2.6) in (6.37) yields the equation

$$\int_0^a \int_0^a \sum_{r=1}^{\infty} \sum_{s=1}^{\infty} \left[\left\{\frac{D\pi^4}{a^4} B_{rs}(r^2+s^2)^2 - q_{rs}\right\} \sin\frac{r\pi x}{a} \sin\frac{s\pi y}{a} - \frac{2\pi^2}{a^2} N_{xy} B_{rs} rs \cos\frac{r\pi x}{a} \cos\frac{s\pi y}{a}\right] \sin\frac{m\pi x}{a} \sin\frac{n\pi y}{a}\, dx\, dy = 0 \tag{6.39}$$

which may be integrated to give

$$\left.\begin{aligned} &B_{mn}(m^2+n^2)^2 - \frac{a^4 q_{mn}}{D\pi^4} \\ &\quad - \frac{32a^2 N_{xy}}{D\pi^4} \sum_{r=1}^{\infty} \sum_{s=1}^{\infty} B_{rs} L(m, r) L(n, s) = 0, \end{aligned}\right\} \tag{6.40}$$

where

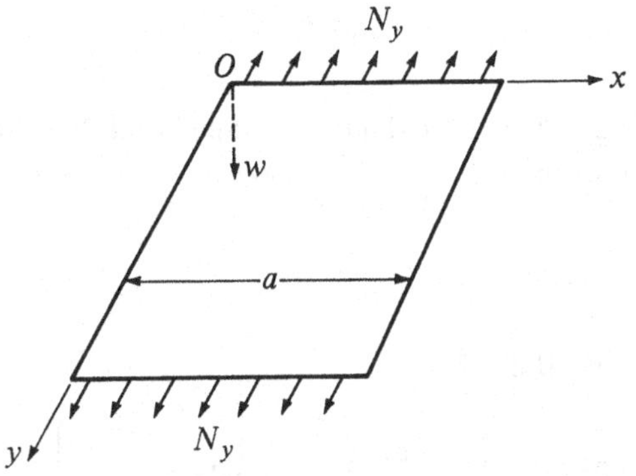

Fig. 6.1

$$L(m, r) = 0, \text{ if } m + r \text{ is an even number;}$$
$$= \frac{mr}{m^2 - r^2}, \text{ if } m + r \text{ is an odd number.}$$

Equations (6.40) are sufficient for determining the coefficients B_{mn}, and hence the deflexion, for any distribution of normal loading and for any given value of N_{xy}. The onset of buckling may likewise be determined from (6.40) by taking $q_{mn} = 0$ and equating to zero the determinant of the coefficients B_{mn}.

6.5 A variational method

The following method was developed independently by Kantorovich (1933), Schurch (1950) and Reissner and Stein (1951). The assumption is made that the deflexion may be expressed in the form

$$w = w_1(x) f_1(y)$$

or, more generally, (6.41)

$$w = \sum_{n=1}^{N} w_n(x) f_n(y)$$

where the $w_n(x)$ are defined functions of x and the $f_n(y)$ are to be determined by variational methods from the condition that Π is a minimum. The method is especially useful when one term alone can be expected to give a reasonable answer. For example, such a case would be the uniformly loaded, clamped, rectangular plate under tensile forces N_y.

6.5.1 Clamped rectangular plate under forces N_y (Fig. 6.1)

If the plate were an infinite strip of width a the deflexion would be given simply by

$$w = \frac{q}{24D} x^2 (a - x)^2 \tag{6.42}$$

and a reasonable answer can be expected if it is assumed that the deflexion of the rectangle is of the form

$$w = \frac{q}{24D} x^2 (a - x)^2 f(y). \tag{6.43}$$

Substitution of (6.43) in (6.14) gives

$$\left.\begin{aligned} \Pi &= \int_0^a \int_0^b \left\{ \tfrac{1}{2} D (\nabla^2 w)^2 + \tfrac{1}{2} N_y \left(\frac{\partial w}{\partial y} \right)^2 - qw \right\} \mathrm{d}x\, \mathrm{d}y \\ &= \frac{q^2 a^5}{11{,}520D} \int_0^b F\, \mathrm{d}y, \\ \text{where} \quad & \\ F &= 8f^2 - \frac{8a^2}{21} f \frac{\mathrm{d}^2 f}{\mathrm{d}y^2} + \frac{a^4}{63} \left(\frac{\mathrm{d}^2 f}{\mathrm{d}y^2} \right) + \frac{a^4 N_y}{63D} \left(\frac{\mathrm{d}f}{\mathrm{d}y} \right)^2 - 16f. \end{aligned}\right\} \tag{6.44}$$

The condition that Π, and hence $\int_0^b F\,\mathrm{d}y$, is a minimum is a result of the *calculus of variations* and, for the general case in which

$$F = F\left(y, f, \frac{\mathrm{d}f}{\mathrm{d}y}, \frac{\mathrm{d}^2 f}{\mathrm{d}y^2}, \ldots, \frac{\mathrm{d}^n f}{\mathrm{d}y^n} \right),$$

is given by *Euler's equation*:

$$\frac{\partial F}{\partial f} - \frac{\mathrm{d}}{\mathrm{d}y}\left(\frac{\partial F}{\partial f_1} \right) + \frac{\mathrm{d}^2}{\mathrm{d}y^2}\left(\frac{\partial F}{\partial f_2} \right) - \cdots + (-1)^n \frac{\mathrm{d}^n}{\mathrm{d}y^n}\left(\frac{\partial F}{\partial f_n} \right) = 0 \tag{6.45}$$

in which the partial derivatives of F are obtained by formally regarding F as a function of independent variables f, $f_1, \ldots,$ where $f_1 = \mathrm{d}f/\mathrm{d}x$, $f_2 = \mathrm{d}^2 f/\mathrm{d}x^2$, and so on.

Substitution of (6.44) in (6.45) yields the equation

$$f - \frac{a^2}{21}\left(1 + \frac{a^2 N_y}{24D} \right) \frac{\mathrm{d}^2 f}{\mathrm{d}y^2} + \frac{a^4}{504} \frac{\mathrm{d}^4 f}{\mathrm{d}y^4} = 1 \tag{6.46}$$

which may be integrated by standard methods.

If N in (6.41) is taken equal to 2, the expression for F is a function of two independent functions f and g, say, and in addition to (6.45) there is

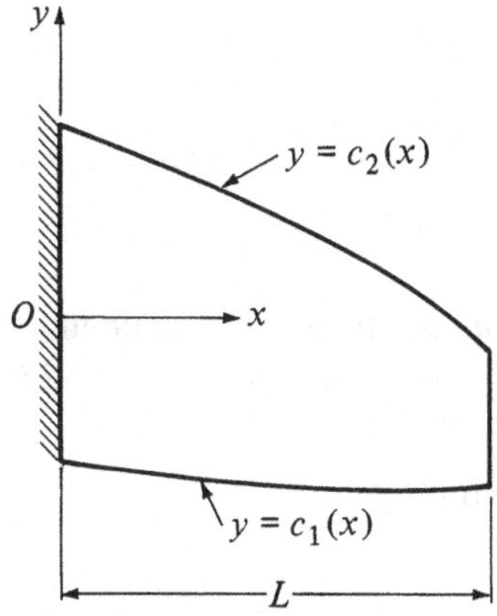

Fig. 6.2

the further equation

$$\frac{\partial F}{\partial g} - \frac{d}{dy}\left(\frac{\partial F}{\partial g_1}\right) + \frac{d^2}{dy^2}\left(\frac{\partial F}{\partial g_2}\right) - \cdots + (-1)^n \frac{d^n}{dy^n}\left(\frac{\partial F}{\partial g_n}\right) = 0, \tag{6.47}$$

and so on for increased values of N.

6.5.2 *Application of variational method to cantilever plates*

Reissner and Stein (1951) applied the variational method to problems of deflexion, vibration and stability of cantilever plates of variable rigidity (Fig. 6.2). The deflexion is assumed to be of the form

$$w = \bar{w}(x) + y\theta(x) \tag{6.48}$$

together with additional terms proportional to higher powers of y, if desired. Here, attention is confined to the determination of the deflexion under a varying pressure $q(x, y)$. Substitution of (6.48) in (6.4) and (6.11) then yields

$$\Pi = U_b + \Pi_q = \frac{1}{2}\int_0^l \left\{ a_1\left(\frac{d^2\bar{w}}{dx^2}\right)^2 + 2a_2\frac{d^2\bar{w}}{dx^2}\frac{d^2\theta}{dx^2} + a_3\left(\frac{d^2\theta}{dx^2}\right)^2 + 2(1-\nu)a_1\left(\frac{d\theta}{dx}\right)^2 \right\} dx - \int_0^l (p_1\bar{w} + p_2\theta)\, dx \tag{6.49}$$

where

$$\left.\begin{aligned} a_n &= \int_{c_1(x)}^{c_2(x)} y^{n-1} D \, \mathrm{d}y \\ p_n &= \int_{c_1(x)}^{c_2(x)} y^{n-1} q \, \mathrm{d}y. \end{aligned}\right\} \tag{6.50}$$

When the variational condition is imposed that Π is a minimum, there results the following simultaneous differential equations for $\bar{w}$ and θ:

$$\frac{\mathrm{d}^2}{\mathrm{d}x^2}\left(a_1 \frac{\mathrm{d}^2\bar{w}}{\mathrm{d}x^2} + a_2 \frac{\mathrm{d}^2\theta}{\mathrm{d}x^2}\right) - p_1 = 0, \tag{6.51}$$

$$\frac{\mathrm{d}^2}{\mathrm{d}x^2}\left(a_2 \frac{\mathrm{d}^2\bar{w}}{\mathrm{d}x^2} + a_3 \frac{\mathrm{d}^2\theta}{\mathrm{d}x^2}\right) - 2(1-\nu)\frac{\mathrm{d}}{\mathrm{d}x}\left(a_1 \frac{\mathrm{d}\theta}{\mathrm{d}x}\right) - p_2 = 0. \tag{6.52}$$

It can likewise be shown that if the cantilever is clamped at the edge $x = 0$ and free along the rest of the boundary, the boundary conditions are

$$\left[\bar{w} = \frac{\mathrm{d}\bar{w}}{\mathrm{d}x} = \theta = \frac{\mathrm{d}\theta}{\mathrm{d}x}\right]_{x=0} = 0 \tag{6.53}$$

and

$$\left.\begin{aligned} \left[a_1 \frac{\mathrm{d}^2\bar{w}}{\mathrm{d}x^2} + a_2 \frac{\mathrm{d}^2\theta}{\mathrm{d}x^2}\right]_{x=l} &= 0 \\ \left[\frac{\mathrm{d}}{\mathrm{d}x}\left(a_1 \frac{\mathrm{d}^2\bar{w}}{\mathrm{d}x^2} + a_2 \frac{\mathrm{d}^2\theta}{\mathrm{d}x^2}\right)\right]_{x=l} &= 0 \\ \left[a_2 \frac{\mathrm{d}^2\bar{w}}{\mathrm{d}x^2} + a_3 \frac{\mathrm{d}^2\theta}{\mathrm{d}x^2}\right]_{x=l} &= 0 \\ \left[\frac{\mathrm{d}}{\mathrm{d}x}\left(a_2 \frac{\mathrm{d}^2\bar{w}}{\mathrm{d}x^2} + a_3 \frac{\mathrm{d}^2\theta}{\mathrm{d}x^2}\right) - 2(1-\nu)a_1 \frac{\mathrm{d}\theta}{\mathrm{d}x}\right]_{x=l} &= 0. \end{aligned}\right\} \tag{6.54}$$

The approximate solution of a cantilever plate problem is now reduced to the solution of (6.51) and (6.52) subject to the boundary conditions (6.53) and (6.54). When the plate (Fig. 6.3) is symmetrical about the x-axis the coefficient a_2 vanishes and the differential equations for $\bar{w}$ and θ become uncoupled:

$$\frac{\mathrm{d}^2}{\mathrm{d}x^2}\left(a_1 \frac{\mathrm{d}^2\bar{w}}{\mathrm{d}x^2}\right) - p_1 = 0 \tag{6.55}$$

and

$$\frac{\mathrm{d}^2}{\mathrm{d}x^2}\left(a_3 \frac{\mathrm{d}^2\theta}{\mathrm{d}x^2}\right) - 2(1-\nu)\frac{\mathrm{d}}{\mathrm{d}x}\left(a_1 \frac{\mathrm{d}\theta}{\mathrm{d}x}\right) - p_2 = 0,$$

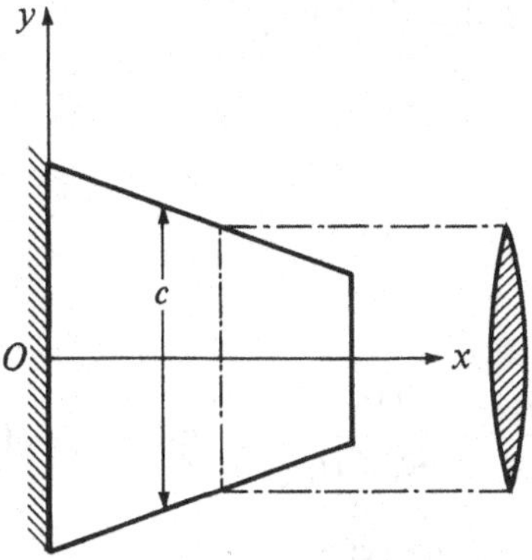

Fig. 6.3

which may be integrated once to give

$$\frac{d}{dx}\left(a_3\frac{d^2\theta}{dx^2}\right)-2(1-\nu)a_1\frac{d\theta}{dx}=-\int_x^l p_2\,dx, \tag{6.56}$$

where the integral is the torque at the section x.

Equation (6.55) can be identified with the flexural equation of a beam of variable rigidity; equation (6.56) cannot be so readily identified because it inherently includes the effect of constraint against axial warping in torsion – an effect overlooked by elementary torsion theory.

Reissner and Stein have shown that closed form solutions of (6.56) may be readily obtained when the rigidity D and the chord c vary according to the laws

$$\left.\begin{aligned} D&=D_0\left(1-\frac{\alpha x}{l}\right)^i K(y/c)\\ c&=c_0\left(1-\frac{\alpha x}{l}\right)^j \end{aligned}\right\} \tag{6.57}$$

where D_0 is the rigidity at the origin and $K(y/c)$ is a symmetrical function of y/c. Solutions for constant chord and exponential rigidity variation of the form $D=D_0e^{-\alpha x/l}$ are also possible.

Linearly varying chord and thickness. As an example, consider the cantilever plate with linearly varying chord and thickness, and lenticular parabolic section, for which

$$\left.\begin{aligned} D&=D_0\left(1-\frac{\alpha x}{l}\right)^3\left\{1-\left(\frac{2y}{c}\right)^2\right\}^3\\ c&=c_0\left(1-\frac{\alpha x}{l}\right) \end{aligned}\right\} \tag{6.58}$$

Substitution of (6.58) in (6.50) and (6.56) and writing

$$\xi = 1 - \alpha x/l$$

$$\varphi = \frac{d\theta}{dx}$$

yields the equation

$$\frac{d}{d\xi}\left(\xi^6 \frac{d\varphi}{d\xi}\right) - \frac{72(1-\nu)l^2\xi^4\varphi}{\alpha^2 c_0^2} = -\frac{315 l^4}{4c_0^3\alpha^4 D_0}\int_\xi^{1-\alpha} p_2(\xi)d\xi \qquad (6.59)$$

which has a complementary function of the type

$$\varphi = A_1\xi^{\beta_1} + A_2\xi^{\beta_2}.$$

Particular integrals may likewise be obtained when p_2 varies as ξ^r, or when a concentrated torque is applied at the tip.

6.6 Variational methods for clamped plates

It was shown in Section 1.4.1 that for plates of constant rigidity the governing biharmonic equation can be expressed as two simultaneous harmonic equations, namely

$$\nabla^2 M = -q \qquad (6.60)$$

and

$$\nabla^2 w = -M/D. \qquad (6.61)$$

The first of these harmonic equations was used by Wegner (1942) in conjunction with a variational procedure for determining the moment term M in plates whose boundaries are clamped, while Morley (1963, 1966) generalized the variational procedure to plates whose boundaries satisfy more general conditions, and he also derived a further variational procedure for determining the deflexion w once M is known.

Among the states of displacement w *which satisfy* (6.60) *and* (6.61), the actual state is determined by a variational equation of least work which minimizes the work done by infinitesimal changes of the bending moments and normal shearing forces that act at the boundary. This work can be equated to the change in the strain energy to yield the variational equation

$$\frac{D}{2}\delta\iint\{(\nabla^2 w)^2 - (1-\nu)\Diamond^4(w,w)\}\,dx\,dy = 0. \qquad (6.62)$$

Further, for the clamped plate of constant rigidity, whose strain energy is given by (6.6), equations (6.61) and (6.62) can be combined to yield the variational equation

$$U_b = \frac{1}{2D}\iint M^2\,dx\,dy = \min, \tag{6.63}$$

where M satisfies (6.60).

Now if M_0 is a particular integral of (6.60), the general solution can be expressed in the form

$$M = M_0 + M_c, \tag{6.64}$$

where M_c satisfies the harmonic equation

$$\nabla^2 M_c = 0. \tag{6.65}$$

The function M_c, and hence M, is to be determined from the condition

$$\iint (M_0 + M_c)^2\,dx\,dy = \min. \tag{6.66}$$

Note that because of the clamped boundary conditions the moment M_n normal to the boundary, where the peak stresses often occur, is given simply by

$$M_n = (1+\nu)M. \tag{6.67}$$

Elsewhere, the values of individual moments M_x and the like require the further integration of (6.61) which can be expressed formally as

$$w = w_0 + w_c, \tag{6.68}$$

where w_0 is a particular integral and w_c satisfies the harmonic equation

$$\nabla^2 w_c = 0. \tag{6.69}$$

Morley (1963) showed that w_c could also be determined from (6.62) and (6.68) which yield the variational equation

$$\iint \Diamond^4(w, \delta w)\,dx\,dy = 0. \tag{6.70}$$

6.6.1 *The uniformly loaded clamped sector*

As a simple example in the application of this method, we outline below the solution for the clamped sectorial plate bounded by the lines $\theta = \pm\alpha$ and unit radius. Because the deflexion is symmetrical about the line $\theta = 0$, we write

$$M = -\frac{q_0 r^2}{4} + q_0 \sum_{m=1}^{\infty} A_m r^{m-1}\cos(m-1)\theta, \tag{6.71}$$

where the first term is a particular integral of (6.60) and the summation satisfies (6.65). Now, in polar coordinates, equation (6.66) is

$$\int_0^1 \int_{-\alpha}^{\alpha} M\delta M r \, dr \, d\theta = 0, \tag{6.72}$$

and this gives rise to the following integrals and associated notation,

$$\begin{aligned} U_{mn} = U_{nm} &= \int_0^1 \int_{-\alpha}^{\alpha} r^{m+n-1} \cos(m-1)\theta \cos(n-1)\theta \, dr \, d\theta \\ &= \frac{1}{m+n}\left\{\frac{\sin(m-n)\alpha}{m-n} + \frac{\sin(m+n-2)\alpha}{m+n-2}\right\} \end{aligned} \tag{6.73}$$

for $m, n = 1, 2, 3\ldots$ and

$$\begin{aligned} U_{m0} &= \frac{1}{4}\int_0^1 \int_{-\alpha}^{\alpha} r^{m+2} \cos(m-1)\theta \, dr \, d\theta \\ &= \frac{\sin(m-1)\alpha}{2(m+3)(m-1)} \end{aligned} \tag{6.74}$$

for $m = 1, 2, 3, \ldots$; the special values for U_{mm}, U_{10} can be derived by the usual limiting processes. When (6.71) is substituted into (6.72) we obtain the following infinite system of ordinary simultaneous equations, symmetrical about the leading diagonal, for the determination of the coefficients A_m

$$\begin{bmatrix} U_{11} & U_{12} & U_{13} & \cdots \\ U_{21} & U_{22} & U_{23} & \cdots \\ U_{31} & U_{32} & U_{33} & \cdots \\ \cdot & \cdot & \cdot & \cdot \end{bmatrix} \begin{bmatrix} A_1 \\ A_2 \\ A_3 \\ \cdots \end{bmatrix} = \begin{bmatrix} U_{10} \\ U_{20} \\ U_{30} \\ \cdots \end{bmatrix}, \tag{6.75}$$

where the system is limited in practice by truncating the infinite series.

Turning attention now to the deflexion w, we express (6.68) in the form

$$w = w_0(r, \theta) + \frac{q_0}{D}\sum_{m=1}^{\infty} B_m r^{m+1} \cos(m+1)\theta, \tag{6.76}$$

where the particular integral, which satisfies (6.61), is taken as

$$w_0 = \frac{q_0 r^4}{64D} - \frac{q_0}{D}\sum_{m=1}^{\infty} \frac{A_m}{4m} r^{m+1} \cos(m-1)\theta, \tag{6.77}$$

where the coefficients A_m have already been determined. The coefficients B_m in (6.76) are now determined from the variational equation (6.70) which is written in polar coordinates as

$$\int_0^1 \int_{-\alpha}^{\alpha} \diamondsuit^4(w, \delta w) r \, dr \, d\theta = 0, \tag{6.78}$$

where

$$\diamondsuit^4(w, \delta w) = \frac{\partial^2 w}{\partial r^2}\left(\frac{1}{r}\frac{\partial \delta w}{\partial r} + \frac{1}{r^2}\frac{\partial^2 \delta w}{\partial \theta^2}\right) - 2\frac{\partial}{\partial r}\left(\frac{1}{r}\frac{\partial w}{\partial \theta}\right)\frac{\partial}{\partial r}\left(\frac{1}{r}\frac{\partial \delta w}{\partial \theta}\right)$$
$$+ \frac{\partial^2 \delta w}{\partial r^2}\left(\frac{1}{r}\frac{\partial w}{\partial r} + \frac{1}{r^2}\frac{\partial^2 w}{\partial \theta^2}\right). \tag{6.79}$$

We now introduce the following notation

$$V_{mn} = V_{nm} = \int_0^1 \int_{-\alpha}^{\alpha} \diamondsuit^4\{r^{m+1}\cos(m+1)\theta, r^{n+1}\cos(n+1)\theta\} r \, dr \, d\theta$$
$$= -\frac{4m(m+1)n(n+1)}{(m+n)(m-n)}\sin(m-n)\alpha \tag{6.80}$$

for $m, n = 1, 2, 3, \ldots$ and

$$V_{m0} = -\frac{D}{q_0}\int_0^1 \int_{-\alpha}^{\alpha} \diamondsuit^4\{r^{m+1}\cos(m+1)\theta, w_0\} r \, dr \, d\theta$$
$$= \frac{m}{4(m+3)}\sin(m+1)\alpha - m(m+1)$$
$$\times \sum_{n=1}^{\infty} \frac{A_n(n-1)}{(m+n)(m-n+2)}\sin(m-n+2)\alpha \tag{6.81}$$

for $m = 1, 2, 3, \ldots$, where the special values can be derived by the usual limiting processes. Substitution of (6.76) into (6.78) now yields the following infinite system of ordinary simultaneous equations for the determination of the coefficients B_m,

$$\begin{bmatrix} V_{11} & V_{12} & V_{13} & \cdots \\ V_{21} & V_{22} & V_{23} & \cdots \\ V_{31} & V_{32} & V_{33} & \cdots \\ \cdot & \cdot & \cdot & \cdot \end{bmatrix} \begin{bmatrix} B_1 \\ B_2 \\ B_3 \\ \cdots \end{bmatrix} = \begin{bmatrix} V_{10} \\ V_{20} \\ V_{30} \\ \cdots \end{bmatrix}. \tag{6.82}$$

References

Duncan, W. J. Galerkin's method in mechanics and differential equations. *Aero. Res. Council R. and M.* No. 1798. H.M.S.O. (August 1937).

Hopkins, H. G. The solution of small displacement, stability or vibration problems concerning a flat rectangular panel when the edges are either clamped or simply supported. *Aero. Res. Council R. and M.* No. 2234. H.M.S.O. (June 1945).

Iguchi, S. Die Knickung der vierseitig eingespannten rechteckigen Platte durch Schubkräfte. *Proc. Phys. Math. Soc. Japan*, **20**, pp. 814–32 (October 1938).
Kantorovich, L. V. *Izv. Akad. Nauk. SSSR*, No. 5 (1933).
Morley, L. S. D. Variational reduction of the clamped plate to two successive membrane problems with an application to uniformly loaded sectors. *Quart. J. Mech. Appl. Math.*, **16**, pp. 451–71 (1963).
———. Some variational principles in plate bending problems. *Quart. J. Mech. App. Math.*, **19**, pp. 371–86 (1966).
Pickett, G. Solution of rectangular clamped plate with lateral load by generalized energy method. *J. Appl. Mech.*, **6**, pp. A-168–70 (December 1939).
Reissner, E., and Stein, M. Torsion and transverse bending of cantilever plates. *N.A.C.A. Tech. Rep.* No. 2369 (June 1951).
Ritz, W. *Gesammelte Werke.* Soc. Suisse de Physique. Gauthier-Villars, Paris, 1911.
Schurch, H. *Inst. für Flugzeugstatik und Flugzeugbau an der E.T.H., Mitt*, No. 2., Zurich, 1950.
Wegner, U. Ein neues Verfahren zur Berechnung der Spannungen in Scheiben. *ForschArb. Geb. IngWes.*, **13**, p. 144 (1942).
Zienciewicz, O. C. *Finite element method.* 3rd ed. McGraw-Hill, 1977.
Sokolnikoff, I. S. *Mathematical theory of elasticity.* 2nd ed. McGraw-Hill, 1956.
Stein, M., Anderson, J., and Hedgepeth, J. M. Deflection and stress analysis of thin solid wings of arbitrary planform with particular reference to Delta wings. *N.A.C.A. Rep.* No. 1131 (1954).

II

LARGE-DEFLEXION THEORY

7

General equations and some exact solutions

The four basic assumptions of small-deflexion theory are summarized in Section 1.1. The first three of these assumptions are retained in large-deflexion theory, but account is now taken of the middle-surface stresses arising from the straining of the middle surface. Such straining occurs, for instance, whenever the plate deflects into a non-developable surface or when the boundary conditions offer restraint against movement in the plane of the plate. The governing equations for isotropic plates are derived in Section 7.1 and some exact solutions of these equations are given in Sections 7.2–7.7. The governing equations for anisotropic plates are given in Section 7.8.

7.1 Governing differential equations for isotropic plates

The equation of equilibrium for a plate with variable thickness and rigidity was derived in Section 1.5 in terms of the rigidity D, deflexion w, and the middle-surface force function Φ:

$$\nabla^2(D\nabla^2 w) - (1-\nu)\Diamond^4(D, w) + (1+\nu)\nabla^2(D\kappa_T) = q + \Diamond^4(\Phi, w), \tag{7.1}$$

where the term involving κ_T refers to the effect of a temperature gradient through the thickness, as discussed in Section 1.6.1.

The force function Φ is not now regarded as independent of the deflexion, and the differential equation satisfied by Φ may be deduced from the stress strain relations below, which are analogous to (1.34),

$$\left.\begin{aligned} \varepsilon_x^0 &= \frac{\partial u}{\partial x} + \frac{1}{2}\left(\frac{\partial w}{\partial x}\right)^2 = \varepsilon_T + (N_x - \nu N_y)/(Et), \\ \varepsilon_y^0 &= \frac{\partial v}{\partial y} + \frac{1}{2}\left(\frac{\partial w}{\partial y}\right)^2 = \varepsilon_T + (N_y - \nu N_x)/(Et), \\ \varepsilon_{xy}^0 &= \frac{\partial u}{\partial y} + \frac{\partial v}{\partial x} + \frac{\partial w}{\partial x}\frac{\partial w}{\partial y} = N_{xy}/(Gt). \end{aligned}\right\} \tag{7.2}$$

In the above, ε_x^0, ε_y^0 and ε_{xy}^0 are the strains in the mid-surface of the plate, and the term ε_T accounts for the effect of a temperature variation in the plane of the plate. The displacements u, v may be eliminated from the strain relations of (7.2) by virtue of the condition of compatibility (1.35) to yield

$$\frac{\partial^2}{\partial y^2}\varepsilon_x^0 + \frac{\partial^2}{\partial x^2}\varepsilon_y^0 - \frac{\partial^2}{\partial x \partial y}\varepsilon_{xy}^0 + \tfrac{1}{2}\lozenge^4(w, w) = 0, \tag{7.3}$$

where use is made of the identity

$$\lozenge^4(w, w) \equiv 2\left\{\frac{\partial^2 w}{\partial x^2}\frac{\partial^2 w}{\partial y^2} - \left(\frac{\partial^2 w}{\partial x \partial y}\right)^2\right\}.$$

Equation (7.3) may be expressed in terms of Φ by means of (7.2) and (1.33) to give

$$\nabla^2\left(\frac{1}{t}\nabla^2\Phi\right) - (1+\nu)\lozenge^4\left(\frac{1}{t}, \Phi\right) + E\nabla^2\varepsilon_T + \tfrac{1}{2}E\,\lozenge^4(w, w) = 0. \tag{7.4}$$

7.1.1 Plate with initial irregularities

Consider a plate initially free from stress but whose mid-surface is given by the equation

$$z = w_0(x, y).$$

For such a plate the moment–curvature relations are as given in (1.62) with w replaced by $(w - w_0)$, whence

$$\frac{\partial^2}{\partial x^2}(w - w_0) = -\frac{12}{Et^3}(M_x - \nu M_y) - \kappa_T; \quad \text{etc.} \tag{7.5}$$

where w is the final shape of the deflected surface referred to the x, y-plane.

Similarly, the mid-surface strains are given by

$$\left.\begin{aligned}
\varepsilon_x^0 &= \frac{\partial u}{\partial x} + \frac{1}{2}\left(\frac{\partial w}{\partial x}\right)^2 - \frac{1}{2}\left(\frac{\partial w_0}{\partial x}\right)^2, \\
\varepsilon_y^0 &= \frac{\partial v}{\partial y} + \frac{1}{2}\left(\frac{\partial w}{\partial y}\right)^2 - \frac{1}{2}\left(\frac{\partial w_0}{\partial y}\right)^2, \\
\varepsilon_{xy}^0 &= \frac{\partial u}{\partial y} + \frac{\partial v}{\partial x} + \frac{\partial w}{\partial x}\frac{\partial w}{\partial y} - \frac{\partial w_0}{\partial x}\frac{\partial w_0}{\partial y},
\end{aligned}\right\} \tag{7.6}$$

so that the *condition of compatibility* may be expressed in the form

$$\frac{\partial^2 \varepsilon_x^0}{\partial y^2} + \frac{\partial^2 \varepsilon_y^0}{\partial x^2} - \frac{\partial^2 \varepsilon_{xy}^0}{\partial x \partial y} + \tfrac{1}{2}\{\lozenge^4(w, w) - \lozenge^4(w_0, w_0)\} = 0. \tag{7.7}$$

It follows that the governing differential equations, corresponding to (7.1) and (7.4) when w_0 is zero, are given by

$$\nabla^2\{D\nabla^2(w-w_0)\}-(1-\nu)\diamondsuit^4(D,w-w_0)+(1+\nu)\nabla^2(D\kappa_T)$$
$$=q+\diamondsuit^4(\Phi,w) \tag{7.8}$$

and

$$\nabla^2\left(\frac{1}{t}\nabla^2\Phi\right)-(1+\nu)\diamondsuit^4\left(\frac{1}{t},\Phi\right)+E\nabla^2\varepsilon_T$$
$$+\tfrac{1}{2}E\{\diamondsuit^4(w,w)-\diamondsuit^4(w_0,w_0)\}=0. \tag{7.9}$$

The solution of a plate problem within the framework of large-deflexion theory reduces to the solution of (7.1) and (7.4) or (7.8) and (7.9), subject to the appropriate boundary conditions. For an unheated and initially flat plate of constant thickness, (7.1) and (7.3) reduce to the equations first derived by von Kármán (1910):

$$D\nabla^4w=q+\diamondsuit^4(\Phi,w), \tag{7.10}$$

$$\nabla^4\Phi=-\tfrac{1}{2}Et\diamondsuit^4(w,w). \tag{7.11}$$

The large-deflexion equations can seldom be solved exactly, but there are notable exceptions including cases where D, t, q and the middle-surface forces are independent of one of the coordinates thus making the problem one-dimensional, or cases of particular and fortuitous variations of D, t and the applied loading. Some such problems will shortly be considered for they throw light on the behaviour in more complex cases. First, however, we derive non-dimensional versions of (7.10), (7.11) and (7.8), (7.9) because these enable us to relate a known large-deflexion solution for a given plate under a given pattern of loads, to a plate with geometrically similar planform and loading pattern. We also derive an expression for the strain energy of a plate in the large-deflexion régime because this is of value in determining whether states are stable or unstable.

7.1.2 Dimensional analysis

We start by introducing non-dimensional coordinates ξ, η such that

$$\xi=x/L, \quad \eta=y/L,$$

where L is a typical planar dimension, such as the width or length. We also write

$$\underset{\sim}{\nabla}^2=\frac{\partial^2}{\partial\xi^2}+\frac{\partial^2}{\partial\eta^2}, \quad \text{etc.} \tag{7.12}$$

so that we have

$$\left.\begin{aligned}\nabla^4&=L^{-4}\underset{\sim}{\nabla}^4,\\ \diamondsuit^4&=L^{-4}\underset{\sim}{\diamondsuit}^4.\end{aligned}\right\} \tag{7.13}$$

The following non-dimensional terms for the loading, deflexion and force function are now introduced:

$$\left.\begin{aligned} q^* &= \frac{qL^4}{Dt}, \\ w^* &= \frac{w}{t}, \\ \Phi^* &= \frac{\Phi}{D}, \end{aligned}\right\} \tag{7.14}$$

and these lead to a non-dimensional form of the von Kármán equations:

$$\underset{\sim}{\nabla}^4 w^* = q^* + \underset{\sim}{\Diamond}^4(\Phi^*, w^*), \tag{7.15}$$

$$\underset{\sim}{\nabla}^4 \Phi^* = -6(1-\nu^2)\underset{\sim}{\Diamond}^4(w^*, w^*). \tag{7.16}$$

Note that it is possible to redefine the terms in (7.14) to remove the $(1-\nu^2)$ factor in (7.16), but only at the expense of greater complexity in the definition of q^*, and so on; furthermore, the advantages to be gained from this are largely illusory because, in general, ν also occurs directly or indirectly through the boundary conditions. Thus, strictly speaking, given solutions of the large-deflexion plate equations can be generalized only to plates with the same value of the Poisson ratio.

Displacement equations

In normally loaded plates with rigid boundaries the in-plane boundary conditions are given by the vanishing of the displacements u, v rather than a specification of the middle-surface forces. In such cases there may be advantages in expressing the large-deflexion equations in terms of the displacements u, v, w. This may be achieved by the elimination of N_x, N_y, N_{xy} from (7.2), (1.32) and (7.10) to give, for the unheated and initially flat plate of constant thickness:

$$\begin{aligned} \frac{t^2}{12}\left(\nabla^4 w - \frac{q}{D}\right) &= \frac{\partial^2 w}{\partial x^2}\left\{\frac{\partial u}{\partial x} + \nu\frac{\partial v}{\partial y} + \frac{1}{2}\left(\frac{\partial w}{\partial x}\right)^2 + \tfrac{1}{2}\nu\left(\frac{\partial w}{\partial y}\right)^2\right\} \\ &\quad + \frac{\partial^2 w}{\partial y^2}\left\{\frac{\partial v}{\partial y} + \nu\frac{\partial u}{\partial x} + \frac{1}{2}\left(\frac{\partial w}{\partial y}\right)^2 + \tfrac{1}{2}\nu\left(\frac{\partial w}{\partial x}\right)^2\right\} \\ &\quad + (1-\nu)\frac{\partial^2 w}{\partial x \partial y}\left(\frac{\partial u}{\partial y} + \frac{\partial v}{\partial x} + \frac{\partial w}{\partial x}\frac{\partial w}{\partial y}\right) \end{aligned} \tag{7.17}$$

and

$$\left.\begin{aligned}
&\frac{\partial}{\partial x}\left[\frac{\partial u}{\partial x}+\frac{\partial v}{\partial y}+\frac{1}{2}\left\{\left(\frac{\partial w}{\partial x}\right)^2+\left(\frac{\partial w}{\partial y}\right)^2\right\}\right]\\
&\qquad+\left(\frac{1-\nu}{1+\nu}\right)\left(\nabla^2 u+\frac{\partial w}{\partial x}\nabla^2 w\right)=0\\
&\frac{\partial}{\partial y}\left[\frac{\partial u}{\partial x}+\frac{\partial v}{\partial y}+\frac{1}{2}\left\{\left(\frac{\partial w}{\partial x}\right)^2+\left(\frac{\partial w}{\partial y}\right)^2\right\}\right]\\
&\qquad+\left(\frac{1-\nu}{1+\nu}\right)\left(\nabla^2 v+\frac{\partial w}{\partial y}\nabla^2 w\right)=0.
\end{aligned}\right\}\tag{7.18}$$

These may also be expressed in non-dimensional form by the further introduction of

$$(u^*, v^*)=\frac{L}{t^2}(u, v)\tag{7.19}$$

to give

$$\begin{aligned}
&\frac{1}{12}(\underset{\sim}{\nabla}^4 w^*-q^*)\\
&\quad=\frac{\partial^2 w^*}{\partial\xi^2}\left\{\frac{\partial u^*}{\partial\xi}+\nu\frac{\partial v^*}{\partial\eta}+\frac{1}{2}\left(\frac{\partial w^*}{\partial\xi}\right)^2+\tfrac{1}{2}\nu\left(\frac{\partial w^*}{\partial\eta}\right)^2\right\}\\
&\qquad+\frac{\partial^2 w^*}{\partial\eta^2}\left\{\frac{\partial v^*}{\partial\eta}+\nu\frac{\partial u^*}{\partial\xi}+\frac{1}{2}\left(\frac{\partial w^*}{\partial\eta}\right)^2+\tfrac{1}{2}\nu\left(\frac{\partial w^*}{\partial\xi}\right)^2\right\}\\
&\qquad+(1-\nu)\frac{\partial^2 w^*}{\partial\xi\partial\eta}\left(\frac{\partial u^*}{\partial\eta}+\frac{\partial v^*}{\partial\xi}+\frac{\partial w^*}{\partial\xi}\frac{\partial w^*}{\partial\eta}\right)
\end{aligned}\tag{7.20}$$

and

$$\left.\begin{aligned}
&\frac{\partial}{\partial\xi}\left[\frac{\partial u^*}{\partial\xi}+\frac{\partial v^*}{\partial\eta}+\frac{1}{2}\left\{\left(\frac{\partial w^*}{\partial\xi}\right)^2+\left(\frac{\partial w^*}{\partial\eta}\right)^2\right\}\right]\\
&\qquad+\left(\frac{1-\nu}{1+\nu}\right)\left(\underset{\sim}{\nabla}^2 u^*+\frac{\partial w^*}{\partial\xi}\underset{\sim}{\nabla}^2 w^*\right)=0\\
&\frac{\partial}{\partial\eta}\left[\frac{\partial u^*}{\partial\xi}+\frac{\partial v^*}{\partial\eta}+\frac{1}{2}\left\{\left(\frac{\partial w^*}{\partial\xi}\right)^2+\left(\frac{\partial w^*}{\partial\eta}\right)^2\right\}\right]\\
&\qquad+\left(\frac{1-\nu}{1+\nu}\right)\left(\underset{\sim}{\nabla}^2 v^*+\frac{\partial w^*}{\partial\eta}\underset{\sim}{\nabla}^2 w^*\right)=0.
\end{aligned}\right\}\tag{7.21}$$

Any solution of (7.15), (7.16) or (7.20), (7.21) thus applies to a range of plates with the same Poisson ratio and geometrically similar planform and pattern of loading *provided the boundary conditions are comparable.*

In this connection we note that the clamped, simply supported and free boundary conditions are comparable because they are homogeneous in x, y so that, for example, (1.67), (1.68), (1.70), (1.76) and (1.78) transform into similar equations with w replaced by w^* and $\boldsymbol{n}, s$ replaced by their non-dimensional counterparts $\underset{\sim}{n}, \underset{\sim}{s}$, where

$$(\underset{\sim}{n}, \underset{\sim}{s}) = (\boldsymbol{n}, s)/L.$$

By the same token, if there are no planar forces applied at a boundary

$$N_n = 0 \quad \text{and} \quad N_{ns} = 0,$$

and hence the non-dimensional boundary conditions are again comparable because they are given by

$$\frac{\partial^2 \Phi^*}{\partial \underset{\sim}{t}^2} = 0 \quad \text{and} \quad \frac{\partial^2 \Phi^*}{\partial \underset{\sim}{n} \partial \underset{\sim}{t}} = 0.$$

Similarly, if there are no planar displacements u, v at a boundary, the non-dimensional boundary conditions are again comparable because

$$u^* = 0 \quad \text{and} \quad v^* = 0.$$

When the boundary conditions are not homogeneous in x, y, solutions may again be applied to other plates subject to the satisfaction of further comparability conditions. For example, at an edge elastically restrained against rotation the boundary condition (1.55) is comparable for plates identified by suffixes 1, 2 if

$$\frac{\chi_1 L_1}{D_1} = \frac{\chi_2 L_2}{D_2}.$$

The above results are not only significant from a theoretical viewpoint, but they are of value in the presentation and condensation of experimental results relating to plates with different thicknesses, overall sizes or elastic moduli. An example is given in Section 9.3.9, which relates to a plate subjected to concentrated loads P. Such loads are limiting cases of distributed loads acting over a small area, and the corresponding non-dimensional term is given by

$$P^* = \frac{PL^2}{Dt}. \tag{7.22}$$

By the same token, the application of a moment M or a torque T to a plate or long strip gives rise to the following non-dimensional terms

$$(M^*, T^*) = \frac{L}{Dt}(M, T). \tag{7.23}$$

Non-dimensional forms of (7.8), (7.9)
For plates with varying thickness and initial irregularities and/or subjected to temperature variations, we introduce the following non-dimensional terms in which t_0, D_0 are reference values of t, D,

$$\left.\begin{aligned} t^* &= t/t_0, \quad D^* = D/D_0, \quad w^* = w/t_0, \quad w_0^* = w_0/t_0, \\ q^* &= \frac{qL^4}{D_0 t_0}, \quad \varepsilon_T^* = \frac{L^2}{t_0^2}, \quad \kappa_T^* = \frac{\kappa_T L^2}{t_0}, \quad \Phi^* = \Phi/D_0. \end{aligned}\right\} \tag{7.24}$$

Equations (7.8), (7.9) can now be expressed in the following non-dimensional form

$$\begin{aligned} &\underset{\sim}{\nabla}^2\{D^*\underset{\sim}{\nabla}^2(w^* - w_0^*)\} - (1-\nu)\underset{\sim}{\Diamond}^4(D^*, w^* - w_0^*) \\ &\quad + (1+\nu)\underset{\sim}{\nabla}^2(D^*\kappa_T^*) = q^* + \underset{\sim}{\Diamond}^4(\Phi^*, w^*) \end{aligned} \tag{7.25}$$

and

$$\begin{aligned} &\underset{\sim}{\nabla}^2\left(\frac{1}{t^*}\underset{\sim}{\nabla}^2\Phi^*\right) - (1+\nu)\,\underset{\sim}{\Diamond}^4\left(\frac{1}{t^*}, \Phi^*\right) + 6(1-\nu^2) \\ &\quad \times \{2\underset{\sim}{\nabla}^2\varepsilon_T^* + \Diamond^4(w^*, w^*) - \underset{\sim}{\Diamond}^4(w_0^*, w_0^*)\} = 0. \end{aligned} \tag{7.26}$$

7.1.3 Strain energy
The strain energy per unit area of plate, U', say, is most readily expressed as the sum of that due to bending stresses, U_b', say, and that due to mid-surface stresses, U_Φ', say. The total strain energy U is then given by

$$\begin{aligned} U &= \iint U' \,\mathrm{d}A \\ &= \iint (U_b' + U_\Phi')\,\mathrm{d}A. \end{aligned} \tag{7.27}$$

Consider first the strain energy due to the moments per unit length in an isotropic plate that may have initial curvatures. In terms of the principal moments M_1, M_2 at any point the strain energy of bending per unit area of deformed plate is given by

$$U_b' = \tfrac{1}{2}(M_1\delta\kappa_1 + M_2\delta\kappa_2), \tag{7.28}$$

where, for example, $\delta\kappa_1$ is the change in the curvature in the direction of the moment M_1; if the plate were initially flat this expression reduces to that in (6.1). Because of the moment–curvature relationships of (1.9) we can write U_b' solely in terms of the principal moments or curvature changes, that is,

$$U_b' = \frac{1}{2D}\{(M_1 + M_2)^2 - 2(1+\nu)M_1M_2\} \tag{7.29}$$

or

$$U_b' = \tfrac{1}{2}D\{(\delta\kappa_1 + \delta\kappa_2)^2 - 2(1-\nu)\delta\kappa_1\,\delta\kappa_2\}. \tag{7.30}$$

For the initially flat plate it follows from Section 1.2.1 that (7.30) can be expressed in the form

$$U_b' = \tfrac{1}{2}D\left[\left(\frac{\partial^2 w}{\partial x^2} + \frac{\partial^2 w}{\partial y^2}\right)^2 - 2(1-\nu)\left\{\frac{\partial^2 w}{\partial x^2}\frac{\partial^2 w}{\partial y^2} - \left(\frac{\partial^2 w}{\partial x \partial y}\right)^2\right\}\right] \tag{7.31}$$

and the invariant nature of this expression becomes apparent if we express it in the form

$$U_b' = \tfrac{1}{2}D\{(\nabla^2 w)^2 - (1-\nu)\Diamond^4(w, w)\}. \tag{7.32}$$

For the plate with initial curvature, the curvature changes are determined by $(w - w_0)$ and hence

$$U_b' = \tfrac{1}{2}D[\{\nabla^2(w - w_0)\}^2 - (1-\nu)\Diamond^4(w - w_0, w - w_0)]. \tag{7.33}$$

The strain energy per unit area due to the middle-surface forces is likewise expressed simply in terms of the principal forces per unit length N_1, N_2. Thus, corresponding to (7.29) we have

$$U_\Phi' = \frac{1}{2Et}\{(N_1 + N_2)^2 - 2(1+\nu)N_1 N_2\}, \tag{7.34}$$

and we note that the suffixes 1, 2 above do not necessarily imply a coincidence between the directions of the principal moments and forces per unit length. If we introduce the force function Φ – see (1.33) – it follows from an analogous argument to that of Section 1.2 that

$$\left.\begin{aligned} N_1 + N_2 &= N_x + N_y, \\ \text{and} \qquad N_1 N_2 &= N_x N_y - N_{xy}^2. \end{aligned}\right\} \tag{7.35}$$

Thus U_Φ' can be expressed in terms of Φ and arbitrary axes in the form

$$U_\Phi' = \frac{1}{2Et}\left[\left(\frac{\partial^2 \Phi}{\partial x^2} + \frac{\partial^2 \Phi}{\partial y^2}\right)^2 - 2(1+\nu)\left\{\frac{\partial^2 \Phi}{\partial x^2}\frac{\partial^2 \Phi}{\partial y^2} - \left(\frac{\partial^2 \Phi}{\partial x \partial y}\right)^2\right\}\right] \tag{7.36}$$

and the invariant nature of this expression becomes apparent if we express it in the form

$$U_\Phi' = \frac{1}{2Et}\{(\nabla^2\Phi)^2 - (1+\nu)\Diamond^4(\Phi, \Phi)\}. \tag{7.37}$$

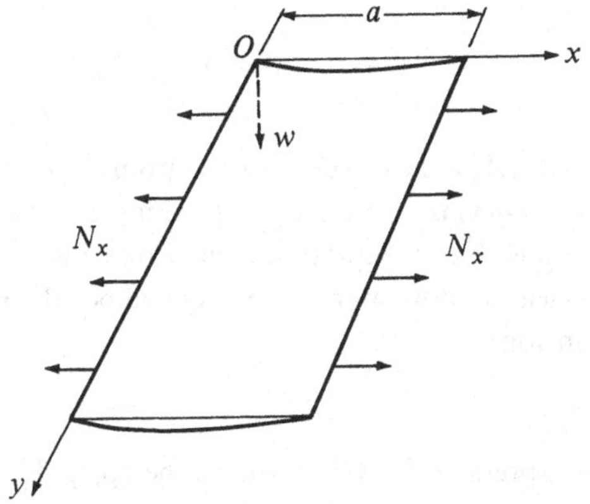

Fig. 7.1

Finally we note that U'_Φ may be expressed in terms of the displacements u, v, w via (7.36) and (7.2) or (7.6).

7.2 Cylindrical deflexion of long strip

The simplest class of one-dimensional problem that admits of solution in the large-deflexion régime is the long strip subjected to a load distribution that does not vary along its length (Fig. 7.1). Such a strip may be treated as a beam. If the edge supports of the strip are free to move in the plane of the plate, there are no middle-surface forces and the small-deflexion solution remains valid. But if the edge supports are rigid, or elastically restrained against movement in the plane of the plate, there is some stretching of the middle surface which gives rise to middle-surface forces N_x. The differential equation for such a strip of constant thickness may be obtained from (7.10):

$$D\frac{\mathrm{d}^4w}{\mathrm{d}x^4} - N_x\frac{\mathrm{d}^2w}{\mathrm{d}x^2} = q(x) = \sum_{m=1}^{\infty} q_m \sin\frac{m\pi x}{a} \qquad (7.38)$$

in which N_x is, at present, unknown.

A particular solution of this equation has already been determined in Section 2.2.3 and the general solution of (7.38) may thus be written in the form

$$\left.\begin{aligned} w = \sum_{m=1}^{\infty} \frac{q_m \sin\dfrac{m\pi x}{a}}{D\left(\dfrac{m\pi}{a}\right)^4 + N_x\left(\dfrac{m\pi}{a}\right)^2} + A_1 + A_2 x \\ + A_3 \sinh\frac{2\eta x}{a} + A_4 \cosh\frac{2\eta x}{a}, \end{aligned}\right\} \qquad (7.39)$$

where

$$\eta^2 = \frac{a^2 N_x}{4D}$$

and the coefficients A_1, A_2, A_3, A_4 are to be determined from the boundary conditions. When the form of $q(x)$ is elementary, for example, uniform or linearly varying, it may be possible to express the particular integral in a simple closed form, but, as will be seen later, there may still be advantages in using the Fourier expansion.

7.2.1 Determination of N_x

Having obtained a formal expression for w in terms of the applied loading and the unknown N_x it is now possible to determine the stretching of the middle surface and thence N_x. First, the degree of elastic restraint of the edges against movement in the plane of the plate must be specified. In many instances, this restraint is provided by regularly spaced stiffeners running across the width of the strip. If the section area of each stiffener is F and their pitch is b, say, a plate tension of N_x per unit length causes a compressive stress in each stiffener equal to bN_x/F. This compressive stress results in the edges of the plate approaching each other by an amount equal to N_x/K, where the edge stiffness K is given by

$$K = \frac{EF}{ab}. \tag{7.40}$$

Now from the first equation of (7.2)

$$\begin{aligned}\frac{N_x}{K} &= -\int_0^a \frac{\mathrm{d}u}{\mathrm{d}x}\,\mathrm{d}x \\ &= \frac{1}{2}\int_0^a \left(\frac{\mathrm{d}w}{\mathrm{d}x}\right)^2 \mathrm{d}x - \frac{\beta a N_x}{Et},\end{aligned} \tag{7.41}$$

where β depends on the longitudinal stiffness of the supporting structure and $1-\nu^2 \leqslant \beta \leqslant 1$, the limits corresponding to the extreme conditions in which $v=0$ and $N_y=0$.

Equations (7.39) and (7.41) are sufficient to determine N_x and thence the deflexion.

Simply supported edges. When the edges are simply supported, the coefficients A_1, A_2, A_3, A_4 in (7.39) are zero, and (7.39) and (7.41) yield the following equation for determing N_x:

$$N_x\left(\frac{1}{K} + \frac{\beta a}{Et}\right) = \frac{a^3}{4\pi^2} \sum_{m=1}^{\infty} \frac{q_m^2}{m^2\left(\dfrac{Dm^2\pi^2}{a^2} + N_x\right)} \tag{7.42}$$

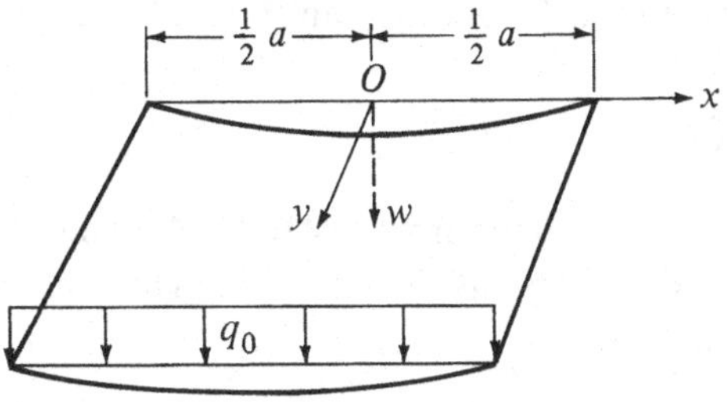

Fig. 7.2

which may be expressed in terms of η:

$$\frac{16\pi^2 D^3}{a^9}\left(\frac{1}{K}+\frac{\beta a}{Et}\right)=\sum_{m=1}^{\infty}\frac{q_m^2}{m^2\eta^2(m^2\pi^2+4\eta^2)^2}. \tag{7.43}$$

Equation (7.43) can be solved by trial and error. The series is very rapidly convergent and a good estimate for N_x may be obtained by considering only the first term.

When the boundary conditions are other than those of simple support, the coefficients A_i in (7.39) are non-zero and a similar analysis for determining N_x in terms of an arbitrary loading $q(x)$ is possible but impracticable. It is preferable to treat each load distribution on its own merits.

Uniform load (Fig. 7.2). If the origin is taken midway between the edges of the strip, and if the boundary conditions at each edge are the same, so that only even powers of x are required, the general solution of (7.38) may be written as

$$w=\frac{q_0 a^4}{8D\eta^2}\left[-\left(\frac{x}{a}\right)^2+A+B\cosh\frac{2\eta x}{a}\right], \tag{7.44}$$

where η is defined in (7.39).

The coefficient B is determined from the relation between the edge moment and slope. If the edges are elastically restrained against rotation so that the boundary conditions are given by (1.72), it is found that

$$\left.\begin{aligned} B&=\frac{1+\frac{1}{2}\lambda}{\lambda\eta\sinh\eta+2\eta^2\cosh\eta},\\ \text{where}\qquad & \\ \lambda&=a\chi/D. \end{aligned}\right\} \tag{7.45}$$

and the limiting cases of simple support and clamping may be obtained from (7.45) by taking $\lambda=0,\ \infty$, respectively.

The coefficient A is determined from the vanishing of w at the edges, so that

$$A = \tfrac{1}{4} - B\cosh\eta. \tag{7.46}$$

The value of N_x may now be found from (7.41), (7.44) and (7.45). It is, in fact, more convenient to regard η as the unknown. Thus we find after integration and rearrangement

$$\frac{256D^3}{q_0^2a^9}\left(\frac{1}{K}+\frac{\beta a}{Et}\right)=\frac{1}{\eta^7}\left\{\frac{\eta}{6}-2B(\eta\cosh\eta-\sinh\eta)\right.$$
$$\left.+B^2\eta^2(\cosh\eta\sinh\eta-\eta)\right\}. \tag{7.47}$$

When the plate is simply supported, the right-hand side of (7.47) reduces to

$$\frac{1}{6\eta^6}-\frac{5}{4\eta^8}+\frac{5\tanh\eta}{4\eta^9}+\frac{\tanh^2\eta}{4\eta^8}.$$

7.3 Uniformly loaded circular plate

The large-deflexion behaviour of a uniformly loaded circular plate will now be discussed from the standpoint of the von Kármán equations. This treatment differs only in detail from that of Way (1934). Because of rotational symmetry, the deflexion w and force function Φ are independent of θ. In expressing (7.10) and (7.11) in polar coordinates, it is convenient to introduce the following non-dimensional parameters augmenting those of (7.14),

$$\left.\begin{aligned}\psi &= \frac{dw^*}{d\rho},\\ \Gamma &= \frac{d\Phi^*}{d\rho},\\ \rho &= r/R,\end{aligned}\right\} \tag{7.48}$$

where R is the radius of the plate.

Substitution of (7.48) in (7.10) and (7.11) gives

$$\left.\begin{aligned}&\frac{d}{d\varrho}\left[\varrho\frac{d}{d\varrho}\left\{\frac{1}{\varrho}\frac{d}{d\varrho}(\varrho\psi)\right\}\right]=q^*\varrho+\frac{d}{d\varrho}(\Gamma\psi)\\ \text{and}\quad&\\ &\frac{d}{d\varrho}\left[\varrho\frac{d}{d\varrho}\left\{\frac{1}{\varrho}\frac{d}{d\varrho}(\varrho\Gamma)\right\}\right]=-12(1-\nu^2)\psi\frac{d\psi}{d\varrho},\end{aligned}\right\} \tag{7.49}$$

and these equations may be integrated once to give

$$\left.\begin{aligned} \varrho\frac{\mathrm{d}}{\mathrm{d}\varrho}\left\{\frac{1}{\varrho}\frac{\mathrm{d}}{\mathrm{d}\varrho}(\varrho\psi)\right\} &= \tfrac{1}{2}q^*\varrho^2+\Gamma\psi \\ \text{and}\qquad \varrho\frac{\mathrm{d}}{\mathrm{d}\varrho}\left\{\frac{1}{\varrho}\frac{\mathrm{d}}{\mathrm{d}\varrho}(\varrho\Gamma)\right\} &= -6(1-\nu^2)\psi^2. \end{aligned}\right\} \tag{7.50}$$

The constants of integration are zero because of the vanishing of ψ, Γ and $\int_0^\varrho q^*\varrho\,\mathrm{d}\varrho$ at the centre.

Equations (7.50) may be solved in series form by assuming

$$\left.\begin{aligned} \varphi &= \sum_{n=1}^{\infty} A_n\varrho^n \\ \Gamma &= \sum_{n=1}^{\infty} B_n\varrho^n. \end{aligned}\right\} \tag{7.51}$$

If (7.51) is substituted in (7.50) and the coefficients of like powers of ϱ equated, it is found that

$$\left.\begin{aligned} B_n &= -6\left(\frac{1-\nu^2}{n^2-1}\right)\sum_{k=1}^{n-2} A_kA_{n-1-k}, \quad n>2 \\ A_n &= \frac{1}{n^2-1}\sum_{k=1}^{n-2} A_kB_{n-1-k}, \quad n>3 \\ A_3 &= (q^*+2A_1B_1)/16. \end{aligned}\right\} \tag{7.52}$$

Equation (7.52) makes it possible to determine the coefficients A_n, B_n in terms of q^*, A_1 and B_1. The coefficients A_1 and B_1 are chosen (possibly by trial and error) to satisfy the boundary conditions. These boundary conditions can be expressed in terms of Γ and ψ as follows:

If there is no edge restraint to movement in the plane of the plate,

$$(\Gamma)_{\varrho=1} = 0; \tag{7.53}$$

if there is no edge displacement in the plane of the plate,

$$\left[\frac{\mathrm{d}\Gamma}{\mathrm{d}\varrho}-\nu\Gamma\right]_{\varrho=1} = 0; \tag{7.54}$$

if the edges of the plate are clamped,

$$(\psi)_{\varrho=1} = 0; \tag{7.55}$$

and if the edges of the plate are simply supported,

$$\left[\frac{\mathrm{d}\psi}{\mathrm{d}\varrho}+\nu\psi\right]_{\varrho=1} = 0. \tag{7.56}$$

A similar analysis is possible whenever the applied loading can be represented by a polynomial in ϱ.

Discussion

The uniformly loaded strip is characterized by the fact that middle-surface forces arise solely from the in-plane stiffness of the supporting structure. Without such support, the middle-surface forces would remain zero, because the deflected form would be a *developable* or *inextensional surface*. In the uniformly loaded circular plate, middle-surface forces arise even if there is no in-plane stiffness of the supporting structure because the plate cannot, for kinematic reasons, deform into a developable surface.

We now consider cases in which the boundary conditions are such that a developable surface is kinematically possible. It will be shown that in the small-deflexion régime the plate deforms, in general, into a non-developable surface but, as the loading is increased into the large-deflexion régime, the action of the induced middle-surface forces is to change the deflected shape into one that can be approximated by a developable surface.

7.4 Pure bending of strip with shallow double curvature

In this section we first derive the governing differential equation for the chordwise distortion of a strip whose thickness t, and hence rigidity D, may vary across the width a. More detailed solutions are then presented for strips whose thickness is constant or varies in a lenticular manner.

The coordinates x, y are measured longitudinally and across the chord, y being zero at the centre line. The initial (constant) curvatures are $\kappa_{x,0}$, $\kappa_{y,0}$ so that in the stress-free state the normal deflexion of the strip is given by

$$w_0 = -\tfrac{1}{2}\kappa_{x,0}x^2 - \tfrac{1}{2}\kappa_{y,0}y^2. \tag{7.57}$$

When the strip is subjected to a moment M, say, the longitudinal curvature becomes κ_x, say, and the deflexion is of the form

$$w = -\tfrac{1}{2}\kappa_x x^2 + \bar{w}(y), \text{ say.} \tag{7.58}$$

The mid-surface forces per unit length are such that N_y and N_{xy} are zero and it is therefore simpler to work directly in terms of N_x rather than the force function Φ. Thus if N_x and the above expressions for w_0 and w are substituted into (7.8) and (7.9) we find after some simplification

$$\frac{d^2}{dy^2}\left\{D\left(\frac{d^2\bar{w}}{dy^2} - \kappa_{y,0} - \nu(\kappa_x - \kappa_{x,0})\right)\right\} = -\kappa_x N_x, \tag{7.59}$$

which is simply the 'beam' equation

$$\frac{d^2M_y}{dy^2} = \kappa_x N_x, \tag{7.60}$$

and

$$\frac{d^2}{dy^2}\left(\frac{N_x}{t}\right) - E\left(\kappa_x \frac{d^2\bar{w}}{dy^2} + \kappa_{x,0}\kappa_{y,0}\right) = 0. \tag{7.61}$$

Equation (7.61) may be integrated, for a given value of κ_x, to give

$$N_x = Et\kappa_x\bar{w} + \tfrac{1}{2}Et\kappa_{x,0}\kappa_{y,0}(y^2 + Ay + B), \tag{7.62}$$

where A, B are constants to be determined from the conditions of longitudinal equilibrium, namely,

$$\int_{-\frac{1}{2}a}^{\frac{1}{2}a} N_x\,dy = 0 \quad \text{and} \quad \int_{-\frac{1}{2}a}^{\frac{1}{2}a} yN_x\,dy = 0. \tag{7.63}$$

Substitution of (7.62) into (7.59) now yields the governing differential equation for $\bar{w}(y)$. The boundary conditions appropriate to this differential equation express the fact that the longitudinal edges are free, whence from (1.74), (1.75) and (7.5),

$$\left.\begin{aligned} &\left[D\left(\frac{d^2\bar{w}}{dy^2} + \kappa_{y,0} - \nu(\kappa_x - \kappa_{x,0})\right)\right]_{y=\pm\frac{1}{2}a} = 0, \\ &\text{and} \\ &\left[\frac{d}{dy}\left\{D\left(\frac{d^2\bar{w}}{dy^2} + \kappa_{y,0} - \nu(\kappa_x - \kappa_{x,0})\right)\right\}\right]_{y=\pm\frac{1}{2}a} = 0. \end{aligned}\right\} \tag{7.64}$$

At this point we note that it is sufficient to determine A, B from (7.63) and only the second term in (7.62), that is,

$$\left.\begin{aligned} &\int_{-\frac{1}{2}a}^{\frac{1}{2}a} t(y^2 + Ay + B)\,dy = 0, \\ &\int_{-\frac{1}{2}a}^{\frac{1}{2}a} t(y^3 + Ay^2 + By)\,dy = 0. \end{aligned}\right\} \tag{7.65}$$

This simplification is possible because integration of (7.59) shows that the vanishing of

$$\int_{-\frac{1}{2}a}^{\frac{1}{2}a} t\bar{w}\,dy \quad \text{and} \quad \int_{-\frac{1}{2}a}^{\frac{1}{2}a} yt\bar{w}\,dy$$

is assured by the boundary conditions (7.64).

7.4.1 The strip of constant thickness

For the strip of constant thickness it follows from (7.65) that

$$A = 0, \quad B = -a^2/12,$$

and hence the differential equation for $\bar{w}$ may be expressed in the form

$$\frac{1}{4k^4}\frac{d^4\bar{w}}{dy^4}+\bar{w}+Q\left(y^2-\frac{a^2}{12}\right)=0, \tag{7.66}$$

where

$$\left.\begin{aligned} k^4 &= \frac{Et\kappa_x^2}{4D},\\ Q &= \frac{\kappa_{x,0}\kappa_{y,0}}{2\kappa_x}. \end{aligned}\right\} \tag{7.67}$$

The solution of (7.66) which satisfies the boundary conditions (7.64) is given by

$$-\bar{w}=Q(y^2-a^2/12)+c_1\cosh ky\cos ky+c_2\sinh ky\sin ky,$$

where

$$\left.\begin{aligned} c_1 &= R\left(\frac{\cos\phi\sinh\phi-\cosh\phi\sin\phi}{\cosh\phi\sinh\phi+\cos\phi\sin\phi}\right),\\ c_2 &= R\left(\frac{\cos\phi\sinh\phi+\cosh\phi\sin\phi}{\cosh\phi\sinh\phi+\cos\phi\sin\phi}\right),\\ R &= \frac{\{\kappa_{y,0}-\nu(\kappa_x-\kappa_{x,0})-2Q\}}{2k^2},\\ \phi &= \tfrac{1}{2}ka. \end{aligned}\right\} \tag{7.68}$$

The total moment applied to the strip contains components due to moments per unit length M_x and middle-surface forces per unit length N_x:

$$M=\int_{-\frac{1}{2}a}^{\frac{1}{2}a}M_x\,dy+\int_{-\frac{1}{2}a}^{\frac{1}{2}a}N_x\bar{w}\,dy. \tag{7.69}$$

Alternatively, we can determine M from the strain energy V per unit length, using the relation

$$M=\frac{dV}{d\kappa_x}. \tag{7.70}$$

First, however, it is convenient to introduce the following non-dimensional expressions for the current and initial curvatures

$$\left.\begin{aligned} &\hat{\kappa}_x=\beta\kappa_x,\quad \hat{\kappa}_{x,0}=\beta\kappa_{x,0},\quad \hat{\kappa}_{y,0}=\beta\kappa_{y,0},\\ &\text{where}\\ &\beta=\frac{a^2\{3(1-\nu^2)\}^{1/2}}{4t}. \end{aligned}\right\} \tag{7.71}$$

Thus we find from Section 7.1.3 after some tedious manipulation

$$V = \frac{2Et^5}{9a^3(1-\nu^2)}\left((\hat{\kappa}_x - \hat{\kappa}_{x,0})^2 + \frac{\lambda^2\Psi_1(\hat{\kappa}_x)}{(1-\nu^2)}\right),$$

where

$$\lambda = (\hat{\kappa}_x - \hat{\kappa}_{x,0})(\nu\hat{\kappa}_x - \hat{\kappa}_{y,0}),$$

$$\Psi_1(\hat{\kappa}_x) = \frac{1}{\hat{\kappa}_x^2}\left[1 - \frac{1}{\hat{\kappa}_x^{1/2}}\left(\frac{\cosh 2\hat{\kappa}_x^{1/2} - \cos 2\hat{\kappa}_x^{1/2}}{\sinh 2\hat{\kappa}_x^{1/2} + \sin 2\hat{\kappa}_x^{1/2}}\right)\right]. \quad (7.72)$$

Equations (7.70) or (7.69) now give

$$M = \frac{Et^4}{3a\{3(1-\nu^2)\}^{1/2}} \times [\hat{\kappa}_x - \hat{\kappa}_{x,0} + \{\lambda/(1-\nu^2)\}\{\mu\Psi_1(\hat{\kappa}_x) - \lambda\hat{\kappa}_x\Psi_2(\hat{\kappa}_x)\}],$$

where

$$\mu = 2\nu\hat{\kappa}_x - \hat{\kappa}_{y,0} - \nu\hat{\kappa}_{x,0},$$

$$\Psi_2(\hat{\kappa}_x) = \frac{1}{\hat{\kappa}_x^4}\left[1 + \frac{\sinh 2\hat{\kappa}_x^{1/2}\sin 2\hat{\kappa}_x^{1/2}}{(\sinh 2\hat{\kappa}_x^{1/2} + \sin 2\hat{\kappa}_x^{1/2})^2} - \frac{5}{4\hat{\kappa}_x^{1/2}}\left(\frac{\cosh 2\hat{\kappa}_x^{1/2} - \cos 2\hat{\kappa}_x^{1/2}}{\sinh 2\hat{\kappa}_x^{1/2} + \sin 2\hat{\kappa}_x^{1/2}}\right)\right] \quad (7.73)$$

The above analysis assumes that there is no torsional component in the deflected shape but, depending on the magnitude of $\kappa_{y,0}$, this may occur during snap-through buckling (for further details, see Mansfield 1973).

A boundary-layer phenomenon

In the expression for $\bar{w}$, see (7.68), we note that as κ_x increases, so too does k, and it follows that for large values of κ_x the hyperbolic terms in $\bar{w}$ mean that near the free edges $\bar{w}$ is an oscillating but rapidly decaying function of distance from the free edges. To investigate this *boundary layer* in greater detail we now consider the pure bending of an initially flat strip for which, from (7.67) and (7.68),

$$Q = 0 \quad \text{and} \quad R = -\nu t\{12(1-\nu^2)\}^{1/2}.$$

If we now introduce

$$Y = \tfrac{1}{2}a - |y|, \quad (7.74)$$

it follows that near a free edge

$$\bar{w} \to \nu t\{12(1-\nu^2)\}^{-1/2}e^{-kY}(\cos kY - \sin kY),$$

as

$$\kappa_x \to \infty, \quad (7.75)$$

and to fix ideas as to the magnitude of $\bar{w}$ we note that for $\nu = \frac{1}{4}$,

$$\bar{w} \to 0.075te^{-kY}(\cos kY - \sin kY).$$

Corresponding to this boundary-layer deflexion, (7.62) shows that

$$N_x = Et\kappa_x \bar{w}, \tag{7.76}$$

and hence from (7.60)

$$\frac{d^2 M_y}{dy^2} = Et\kappa_x^2 \bar{w},$$

$$\to 2\nu D\kappa_x k^2 e^{-kY}(\cos kY - \sin kY), \tag{7.77}$$

by virtue of (7.67), (7.75).

A preliminary integration of (7.77) yields

$$\frac{dM_y}{dY} \to 2\nu D\kappa_x k e^{-kY} \sin kY,$$

and hence

$$M_y \to \nu D\kappa_x \{1 - e^{-kY}(\cos kY + \sin kY)\}. \tag{7.78}$$

The boundary layer thus introduces large but localized values of N_x which cause a rapid build-up in the moment M_y from zero at the edges to $\nu D\kappa_x$ away from the edges. The moment $\nu D\kappa_x$ corresponds to bending with zero transverse curvature κ_y, that is, bending into a developable surface. The term $1/k$ provides a measure of the width of the boundary layer Y_{BL}, and hence

$$Y_{BL} \approx 0.77\left(\frac{t}{\kappa_x}\right)^{1/2}. \tag{7.79}$$

A similar boundary layer occurs in the large-deflexion régime near any free edge where the thickness t is non-zero.

7.4.2 *The strip of lenticular section*

It might be thought that the large-deflexion analysis of an initially curved strip of lenticular section would be more complex than that of the strip of constant thickness, but this is not so. The reason for this relative simplicity stems from the fact that the rigidity tapers smoothly to zero at the boundaries, and this has the effect of ironing out the boundary layers which would otherwise occur.

Consider therefore a strip whose thickness, and hence rigidity, varies as follows

$$\left.\begin{aligned} t &= t_0\{1-(2y/a)^2\}, \\ D &= D_0\{1-(2y/a)^2\}^3, \\ D_0 &= Et_0^3/\{12(1-\nu^2)\}. \end{aligned}\right\} \tag{7.80}$$

When (7.80) is substituted into (7.59), (7.62), (7.64) and (7.65) it may be verified that

$$\left.\begin{aligned} A &= 0, \\ B &= -a^2/20, \\ \bar{w} &= -\tfrac{1}{2}\kappa_y(y^2 - a^2/20), \end{aligned}\right\} \tag{7.81}$$

where κ_y depends on $\kappa_{x,0}$, $\kappa_{y,0}$ and κ_x, but is independent of y so that the curvature does not vary across the chord – a feature which is peculiar to the lenticular section. Before proceeding further, however, it is convenient to reintroduce and *redefine* the following non-dimensional terms:

$$\left.\begin{aligned} \{\hat{\kappa}_{x,0}, \hat{\kappa}_{y,0}, \hat{\kappa}_x, \hat{\kappa}_y\} &= \frac{a^2}{4t_0}\left(\frac{1-\nu^2}{5}\right)^{1/2}\{\kappa_{x,0}, \kappa_{y,0}, \kappa_x, \kappa_y\}, \\ \hat{M} = \left(\frac{21a\{5(1-\nu^2)\}^{1/2}}{16Et_0^4}\right)M, &\quad \hat{V} = \left(\frac{21a^2(1-\nu^2)}{64Et_0^5}\right)V. \end{aligned}\right\} \tag{7.82}$$

In terms of these non-dimensional expressions we now find

$$\left.\begin{aligned} \hat{\kappa}_y &= \frac{\hat{\kappa}_{y,0} - \nu(\hat{\kappa}_x - \hat{\kappa}_{x,0}) + \hat{\kappa}_x\hat{\kappa}_{x,0}\hat{\kappa}_{y,0}}{1+\hat{\kappa}_x^2}, \\ \hat{V} &= \frac{1}{2(1-\nu^2)}[(\hat{\kappa}_x + \hat{\kappa}_y - \hat{\kappa}_{x,0} - \hat{\kappa}_{y,0})^2 \\ &\quad - 2(1-\nu)(\hat{\kappa}_x - \hat{\kappa}_{x,0})(\hat{\kappa}_y - \hat{\kappa}_{y,0}) \\ &\quad + (\hat{\kappa}_{x,0}\hat{\kappa}_{y,0} - \hat{\kappa}_x\hat{\kappa}_y)^2]. \end{aligned}\right\} \tag{7.83}$$

Note that a check on (7.83) is afforded by the fact that

$$\partial\hat{V}/\partial\hat{\kappa}_y = 0. \tag{7.84}$$

The non-dimensional bending moment is now given by

$$\begin{aligned} \hat{M} &= \mathrm{d}\hat{V}/\mathrm{d}\hat{\kappa}_x, \\ &= \left(\frac{\partial\hat{V}}{\partial\hat{\kappa}_x}\right)_{\hat{\kappa}_y\,\text{const.}} \end{aligned}$$

by virtue of (7.84), and hence

$$\left.\begin{aligned}&\hat{M}=\hat{\kappa}_x-\hat{\kappa}_{x,0}+\frac{\lambda}{(1-\nu^2)(1+\hat{\kappa}_x^2)}\left(\mu-\frac{\lambda\hat{\kappa}_x}{1+\hat{\kappa}_x^2}\right)\\ \text{where}\quad&\\ &\lambda=(\hat{\kappa}_x-\hat{\kappa}_{x,0})(\nu\hat{\kappa}_x-\hat{\kappa}_{y,0}),\\ &\mu=2\nu\hat{\kappa}_x-\hat{\kappa}_{y,0}-\nu\hat{\kappa}_{x,0}.\end{aligned}\right\}\tag{7.85}$$

The above analysis again assumes that there is no torsional component in the deflexion but, depending on the magnitude of $\hat{\kappa}_{y,0}$, this may occur during snap-through buckling when the direction of the applied moment is such that destabilizing *compressive* middle-surface forces occur near the edges. Thus the writer showed (1973) that buckling into a torsional mode can occur whenever

$$\left.\begin{aligned}&\hat{\kappa}_{y,0}>(2-\nu)\hat{\kappa}_{x,0}+2\{(1-\nu)(1+\hat{\kappa}_{x,0}^2)\}^{1/2},\\ \text{or}\quad&\hat{\kappa}_{y,0}<(2-\nu)\hat{\kappa}_{x,0}-2\{(1-\nu)(1+\hat{\kappa}_{x,0}^2)\}^{1/2},\end{aligned}\right\}\tag{7.86}$$

and that such buckling occurs when

$$\begin{aligned}\hat{\kappa}_x=\tfrac{1}{2}(\hat{\kappa}_{y,0}+\nu\hat{\kappa}_{x,0})\mp\tfrac{1}{2}\{(\hat{\kappa}_{y,0}+\nu\hat{\kappa}_{x,0})^2\\ -4(1-\nu+\hat{\kappa}_{x,0}\hat{\kappa}_{y,0})\}^{1/2}.\end{aligned}\tag{7.87}$$

Similarly, when the direction of the applied moment is such that there are *tensile* middle-surface forces near the edges, flexural snap-through buckling occurs if

$$d\hat{M}/d\hat{\kappa}_x=0,\tag{7.88}$$

and it follows from (7.85) that flexural snap-through buckling can occur whenever

$$\left.\begin{aligned}&\hat{\kappa}_{y,0}>-(2+\nu)\hat{\kappa}_{x,0}+2\{(1+\nu)(1+\hat{\kappa}_{x,0}^2)\}^{1/2},\\ \text{or}\quad&\hat{\kappa}_{y,0}<-(2+\nu)\hat{\kappa}_{x,0}-2\{(1+\nu)(1+\hat{\kappa}_{x,0}^2)\}^{1/2}.\end{aligned}\right\}\tag{7.89}$$

Finally, we note that for large values of $\hat{\kappa}_x$, (7.83) shows that

$$\left.\begin{aligned}&\hat{\kappa}_y\sim(\hat{\kappa}_{x,0}\hat{\kappa}_{y,0}-\nu)/\hat{\kappa}_x,\\ &\quad\to0\text{ as }\hat{\kappa}_x\to\infty,\end{aligned}\right\}\tag{7.90}$$

so that the strip approximates to a developable surface in which

$$\hat{M}\sim\hat{\kappa}_x/(1-\nu^2),\tag{7.91}$$

regardless of the magnitude of the initial curvatures.

7.5 Flexure and torsion of flat strip of lenticular section

When a long strip is subjected to a combination of pure bending and torsion, the analysis is more complex and, to simplify the discussion, we

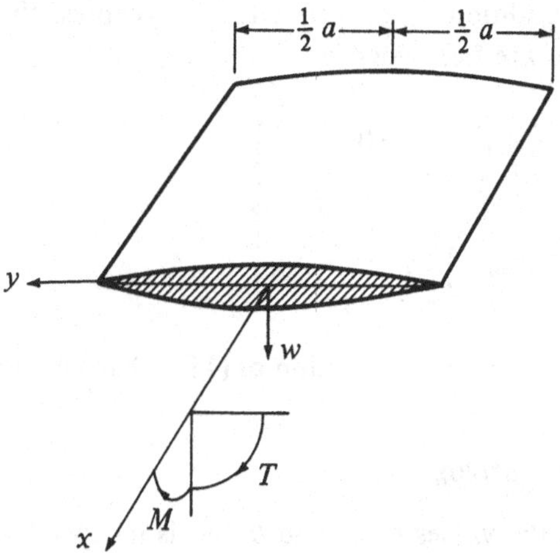

Fig. 7.3

confine attention to the initially flat strip of lenticular section [see (7.80) and Fig. 7.3]. Under the action of a moment M and a torque T, the deflexion of the strip is of the form

$$w = -\tfrac{1}{2}\kappa_x x^2 - \theta xy + \bar{w}(y), \tag{7.92}$$

where θ is the twist per unit length and the other terms are as defined in Section 7.4.

Substitution of (7.92) in (7.8) and (7.9) gives

$$\frac{d^2}{dy^2}\left\{D\left(\frac{d^2\bar{w}}{dy^2} - \nu\kappa_x\right)\right\} = -\kappa_x N_x \tag{7.93}$$

and

$$\frac{d^2}{dy^2}\left(\frac{N_x}{t}\right) - E\left(\theta^2 + \kappa_x\frac{d^2\bar{w}}{dy^2}\right) = 0. \tag{7.94}$$

Equation (7.94) may be integrated for given values of κ_x, θ to give

$$N_x = Et\{\kappa_x\bar{w} + \tfrac{1}{2}\theta^2(y^2 + Ay + B)\}, \tag{7.95}$$

where the constants of integration A, B are to be determined from the conditions of longitudinal equilibrium whence, from (7.65),

$$A = 0, \quad B = -a^2/20. \tag{7.96}$$

Substitution into (7.93) now yields the following differential equation for $\bar{w}$:

$$\frac{d^2}{dy^2}\left\{D\left(\frac{d^2\bar{w}}{dy^2} - \nu\kappa_x\right)\right\} + \kappa_x Et\{\kappa_x\bar{w} + \tfrac{1}{2}\theta^2(y^2 - a^2/20)\}. \tag{7.97}$$

The four boundary conditions appropriate to (7.97) express the fact that the longitudinal edges are free, whence

$$\left.\begin{aligned} &\left[D\left(\frac{\mathrm{d}^2\bar{w}}{\mathrm{d}y^2}-\nu\kappa_x\right)\right]_{y=\pm\frac{1}{2}a}=0 \\ \text{and} \qquad &\left[\frac{\mathrm{d}}{\mathrm{d}y}\left\{D\left(\frac{\mathrm{d}^2\bar{w}}{\mathrm{d}y^2}-\nu\kappa_x\right)\right\}\right]_{y=\pm\frac{1}{2}a}=0. \end{aligned}\right\} \tag{7.98}$$

It may now be verified that the solution of (7.97) that satisfies (7.98) is of the form

$$\bar{w}=-\tfrac{1}{2}\kappa_y(y^2-a^2/20), \tag{7.99}$$

where κ_y depends on the values of κ_x and θ, but is independent of y, as in Section 7.4.2. At this point, however, it is convenient to reintroduce the non-dimensional terms $\hat{\kappa}_x, \hat{\kappa}_y, \hat{M}, \hat{V}$ of (7.82) and augment them with

$$\left.\begin{aligned} \hat{\theta}&=\frac{a^2}{4t_0}\left(\frac{1-\nu^2}{5}\right)^{1/2}\theta, \\ \hat{N}&=\frac{a^2}{2Et_0^3}[N_x]_{y=0}, \\ \hat{T}&=\frac{21a5^{1/2}}{64Gt_0^4}T, \end{aligned}\right\} \tag{7.100}$$

where G is the shear modulus.

In terms of these parameters, we now find

$$\left.\begin{aligned} \hat{\kappa}_y&=-\left(\frac{\nu-(1-\nu^2)\hat{\theta}^2}{1+(1-\nu^2)\hat{\kappa}_x^2}\right)\hat{\kappa}_x, \\ \hat{N}&=\frac{\hat{\theta}^2+\nu\hat{\kappa}_x^2}{1+(1-\nu^2)\hat{\kappa}_x^2}, \\ \hat{V}&=\tfrac{1}{2}\hat{\kappa}_x^2+\frac{\hat{\theta}^2}{1+\nu}+\frac{(\hat{\theta}^2+\nu\hat{\kappa}_x^2)^2}{2\{1+(1-\nu^2)\hat{\kappa}_x^2\}}, \end{aligned}\right\} \tag{7.101}$$

and finally,

$$\left.\begin{aligned} \hat{M}&=\frac{\partial\hat{V}}{\partial\hat{\kappa}_x}=\frac{\hat{\kappa}_x\{1+(1+\nu)(\hat{\kappa}_x^2+\hat{\theta}^2)\}\{1+(1-\nu)(\hat{\kappa}_x^2-\hat{\theta}^2)\}}{\{1+(1-\nu^2)\hat{\kappa}_x^2\}^2} \\ \text{and} \qquad \hat{T}&=\left(\frac{1+\nu}{2}\right)\frac{\partial\hat{V}}{\partial\hat{\theta}}=\frac{\hat{\theta}\{1+(1+\nu)(\hat{\kappa}_x^2+\hat{\theta}^2)\}}{1+(1-\nu^2)\hat{\kappa}_x^2}. \end{aligned}\right\} \tag{7.102}$$

Equations (7.102) are the large-deflexion counterparts of the small-deflexion relations between bending moment and curvature, and between torque and twist per unit length, which may be expressed in terms of $\hat{M}$, $\hat{\kappa}$, $\hat{T}$ and $\hat{\theta}$ as follows:

$$\hat{M} = \hat{\kappa}_x \quad \text{and} \quad \hat{T} = \hat{\theta}.$$

It is seen that the large-deflexion relations (7.102) are not only non-linear but are also coupled, and in the following discussion we first consider the behaviour of the strip under pure moment or pure torque.

7.5.1 Strip under pure moment

The condition that $\hat{T}$ is zero and $\hat{M}$ non-zero implies that $\hat{\theta}$ is zero. Equations (7.101) and (7.102) then yield

$$\left.\begin{aligned} \hat{\kappa}_y &= \frac{-\nu\hat{\kappa}_x}{1+(1-\nu^2)\hat{\kappa}_x^2}, \\ \hat{N} &= \frac{\nu\hat{\kappa}_x^2}{1+(1-\nu^2)\hat{\kappa}_x^2}, \\ \hat{M} &= \frac{\hat{\kappa}_x}{1-\nu^2}\left\{1-\left(\frac{\nu}{1+(1-\nu^2)\hat{\kappa}_x^2}\right)^2\right\}. \end{aligned}\right\} \tag{7.103}$$

These relations show that for large values of $\hat{\kappa}_x$

$$\left.\begin{aligned} &\hat{\kappa}_y \to 0, \\ &\hat{N} \to \frac{\nu}{1-\nu^2}, \\ \text{and} \quad &\hat{M} \to \frac{\hat{\kappa}_x}{1-\nu^2}, \end{aligned}\right\} \tag{7.104}$$

so that the strip tends to a developable surface, and the middle-surface forces approach constant values.

7.5.2 Strip under pure torque

The condition that $\hat{M}$ is zero and $\hat{T}$ non-zero implies that either

$$\left.\begin{aligned} &\hat{\kappa}_x = 0, \\ &\text{in which case} \\ &\hat{\kappa}_y = 0, \\ &\hat{N} = \hat{\theta}^2, \\ &\hat{T} = \hat{\theta}\{1+(1+\nu)\hat{\theta}^2\}; \end{aligned}\right\} \tag{7.105}$$

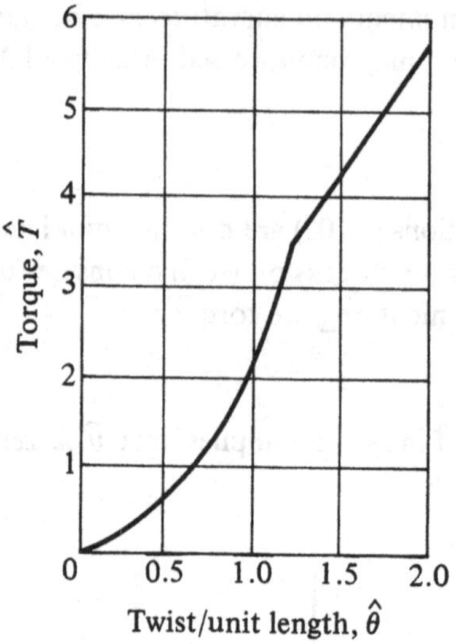

Fig. 7.4

or

$$1+(1-\nu)(\hat{\kappa}_x^2-\hat{\theta}^2)=0,$$

in which case

$$\left.\begin{aligned}\hat{\kappa}_y=\hat{\kappa}_x&=\pm\left(\hat{\theta}^2-\frac{1}{1-\nu}\right)^{1/2},\\ \hat{N}&=\frac{1}{1-\nu},\\ \hat{T}&=\frac{2\hat{\theta}}{1-\nu}.\end{aligned}\right\}\qquad(7.106)$$

It is possible to determine which of these states is the correct one from a comparison of their strain energies. Thus we find that (7.105) is appropriate for values of $|\hat{T}|$ up to a critical value $\hat{T}^*$ where

$$\hat{T}^*=2(1-\nu)^{-3/2}\quad\text{and}\quad\hat{\theta}^*=(1-\nu)^{-1/2},\qquad(7.107)$$

and (7.106) is appropriate for values of $\hat{T}$ greater than $\hat{T}^*$, as shown in Fig. 7.4.

As the torque increases through the critical value of $\hat{T}^*$, the strip buckles into one of two possible, and equally probable, modes characterized by (7.106). This instability is due to the fact that as the strip is twisted, the middle-surface forces play an increasing part in resisting the torque, and eventually the strip buckles and deforms into a surface which approximates to a developable surface, thus halting the increase in the middle-surface strains and concomitant forces.

When $\hat{T}$ is large compared with $\hat{T}^*$, the lateral and longitudinal curvatures are given by

$$\hat{\kappa}_y = \hat{\kappa}_x = \pm\hat{\theta} + O(\hat{\theta}^{-1})$$

which shows that the deflected shape tends to a developable surface whose generators lie at $\pm 45°$ to the longitudinal axis of the strip.

Bending moment and torque increasing in fixed ratio. If M and T increase in proportion such that $M = cT$, say, it follows from (7.82), (7.100) that

$$\hat{M} = \frac{2c}{1+\nu}\hat{T}$$

and the relationship between $\hat{T}$ and $\hat{\theta}$, for example, is found by eliminating $\hat{\kappa}_x$ from (7.102) to give

$$\hat{\theta}^2 = \frac{\{(\hat{T}/\hat{\theta}) - 1\}\{2-(1-\nu)\hat{T}/\hat{\theta}\}^2}{(1+\nu)[4c^2\{1-(1-\nu)\hat{T}/\hat{\theta}\} + \{2-(1-\nu)\hat{T}/\hat{\theta}\}^2]}. \tag{7.108}$$

The variation of $\hat{\kappa}_x, \hat{\kappa}_y$ with $\hat{T}$ (or $\hat{M}$) now follows immediately from (7.102) and (7.101); in particular, it can be shown that for large values of the applied moments the ratios $\hat{T}/\hat{\theta}$, $\hat{T}/\hat{\kappa}_x$, $\hat{T}/\hat{\kappa}_y$ tend to constant values given by

$$\left.\begin{aligned}
\frac{\hat{T}}{\hat{\theta}} &\to \frac{2}{1-\nu}\left(\frac{\sqrt{(1+c^2)}}{c+\sqrt{(1+c^2)}}\right) + O(\hat{T}^{-2})\\
\frac{\hat{T}}{\hat{\kappa}_x} &\to \frac{2}{1-\nu}\left(\frac{\sqrt{(1+c^2)}}{\{c+\sqrt{(1+c^2)}\}^2}\right) + O(\hat{T}^{-2})\\
\frac{\hat{T}}{\hat{\kappa}_y} &\to \frac{1}{1-\nu}\sqrt{(1+c^2)} + O(\hat{T}^{-2}).
\end{aligned}\right\} \tag{7.109}$$

It can be seen from these asymptotic expressions that

$$\hat{\kappa}_x\hat{\kappa}_y - \hat{\theta}^2 \to 0$$

which shows that the surface tends to a developable one. The angle that the generators of this developable surface make with the x-axis is given by

$$\tfrac{1}{2}\cot^{-1}\left(\frac{\hat{\kappa}_x - \hat{\kappa}_y}{2\hat{\theta}}\right) = \tfrac{1}{2}\tan^{-1}\left(\frac{T}{M}\right). \tag{7.110}$$

Chapter 8 shows that this result is in agreement with *inextensional theory*, a simplified large-deflexion theory based on the assumption of an inextensible middle surface for which the only possible mode of deformation is a developable surface.

7.6 Elliptical plate with temperature gradient through the thickness

We continue our presentation of exact solutions of the large-deflexion plate equations with the analysis of an unsupported and initially flat elliptical plate whose thickness varies parabolically across a diameter, vanishing along the boundary according to the equation

$$t = t_0\left(1 - \frac{x^2}{a^2} - \frac{y^2}{b^2}\right), \tag{7.111}$$

where t_0 is the thickness at the centre and a, b are the major and minor semi-axes of the plate. The rigidity D is thus given by

$$\left.\begin{aligned} D &= D_0\left(1 - \frac{x^2}{a^2} - \frac{y^2}{b^2}\right)^3, \\ \text{where} \quad D_0 &= \frac{Et_0^3}{12(1-\nu^2)}. \end{aligned}\right\} \tag{7.112}$$

The temperature distribution is such that the temperature gradient $\partial T/\partial z$ through the thickness is constant, as is the temperature of the mid-plane of the plate. This temperature distribution would cause a uniform 'spherical' curvature change κ_T in each *unrestrained* element of the plate such that

$$\kappa_T = \alpha(\partial T/\partial z), \tag{7.113}$$

where α is the coefficient of thermal expansion. The symbol κ_T thus provides a convenient measure of the magnitude of the temperature gradient.

At this stage it is convenient to anticipate some of the later results by drawing attention to two peculiar features of the heated plate. The first feature is that the deflexion is of the form

$$w = -\tfrac{1}{2}(\kappa_x x^2 + 2\kappa_{xy}xy + \kappa_y y^2), \tag{7.114}$$

where $\kappa_x, \kappa_{xy}, \kappa_y$ depend upon κ_T but are independent of x, y. The deflexion of the plate is thus completely determined by the curvatures $\kappa_x, \kappa_{xy}, \kappa_y$ or, of course, two principal curvatures and their associated angle Ω.

The second feature is that although the *magnitude* of the middle-surface forces depends upon κ_T, their *distribution* does not. The middle-surface forces are derived from a force function Φ that varies in the same manner as does the rigidity D. Furthermore, the rigidity D has the same dimensions (force × length) as the force function, and it is therefore convenient to write

$$\Phi = \beta D, \tag{7.115}$$

where β is non-dimensional.

The analysis and presentation of results is also simplified by the introduction of non-dimensional curvature symbols, and these are identified by a circumflex:

$$\left.\begin{aligned} &\{\hat{\kappa}_T, \hat{\kappa}_x, \hat{\kappa}_y, \hat{\kappa}_{xy}\} = k\{\kappa_T, \kappa_x, \kappa_y, \kappa_{xy}\}, \\ &\text{where} \\ &k = \frac{ab}{t_0}\left(\frac{1-\nu^2}{4+2\nu+5(a^2/b^2+b^2/a^2)}\right)^{1/2}. \end{aligned}\right\} \tag{7.116}$$

Boundary conditions

We have already anticipated the forms that w and Φ assume, and we now show that these satisfy the boundary conditions. There are no forces or moments applied to the edge of the plate, so that along the boundary

$$\left.\begin{aligned} &N_n = N_{ns} = 0, \\ &M_n = Q_n + \partial M_{ns}/\partial_s = 0. \end{aligned}\right\} \tag{7.117}$$

Now the variation of D, and hence Φ, is such that along the boundary

$$D = \frac{\partial D}{\partial n} = \Phi = \frac{\partial \Phi}{\partial n} = 0,$$

and accordingly the boundary conditions are satisfied. It remains to show how the disposable parameters in the expressions for w and Φ, namely $\kappa_x, \kappa_{xy}, \kappa_y$ and β, may be chosen to satisfy the governing partial differential equations. In this sense the method of solution is an inverse one.

7.6.1 *Satisfaction of the governing differential equations*

The satisfaction of (7.1) and (7.4) by expression (7.114) and (7.115) is facilitated by the following general identity

$$\nabla^2 D \equiv \Diamond^4\{D, \tfrac{1}{2}(x^2+y^2)\}, \tag{7.118}$$

and the following equalities peculiar to this particular problem:

$$\left.\begin{aligned} &\nabla^2\left(\frac{1}{t}\nabla^2\Phi\right) = \frac{12\beta D_0}{t_0 a^2 b^2}\left\{2+5\left(\frac{a^2}{b^2}+\frac{b^2}{a^2}\right)\right\}, \text{ a constant,} \\ &\text{and} \\ &\Diamond^4\left(\frac{1}{t}, \Phi\right) = -24\beta D_0/t_0 a^2 b^2, \text{ a constant.} \end{aligned}\right\} \tag{7.119}$$

Thus, referring first to (7.4), with ε_T zero, we find that each term is a constant and the equation can be expressed non-dimensionally as

$$\beta + \hat{\kappa}_x \hat{\kappa}_y - \hat{\kappa}_{xy}^2 = 0. \tag{7.120}$$

Similarly, (7.1), with q zero, may be expressed in the form

$$\left.\begin{aligned} &\lozenge^4(D, \tfrac{1}{2}Ax^2 + Bxy + \tfrac{1}{2}Cy^2) = 0,\\ &\text{where}\\ &A = (1+\nu)\hat{\kappa}_T + (\beta - \nu)\hat{\kappa}_x - \hat{\kappa}_y,\\ &B = (1 + \beta - \nu)\hat{\kappa}_{xy},\\ &C = (1+\nu)\hat{\kappa}_T + (\beta - \nu)\hat{\kappa}_y - \hat{\kappa}_x, \end{aligned}\right\} \tag{7.121}$$

and the satisfaction of (7.121) is assured by the vanishing of A, B and C. The resulting three equations, together with (7.120), constitute four equations for determining $\hat{\kappa}_x$, $\hat{\kappa}_y$, $\hat{\kappa}_{xy}$ and β. In particular, the vanishing of B implies that either

$$\hat{\kappa}_{xy} = 0, \tag{7.122}$$

or

$$\beta = -(1 - \nu), \tag{7.123}$$

and we first consider (7.122) because this includes the solution for small values of $\hat{\kappa}_T$.

Equations (7.120) and (7.121) now yield

$$\begin{aligned} \hat{\kappa}_x = \hat{\kappa}_y &= \hat{\kappa}, \quad \text{say,}\\ &= \left(\frac{1+\nu}{1+\nu-\beta}\right)\hat{\kappa}_T, \end{aligned} \tag{7.124}$$

where β is the real root of the cubic

$$\beta(1+\nu-\beta)^2 + (1+\nu)^2\hat{\kappa}_T^2 = 0. \tag{7.125}$$

For small values of $\hat{\kappa}_T$ the plate thus deforms into a spherical surface and we note the following simple variation of β with $\hat{\kappa}$,

$$\beta = -\hat{\kappa}^2,$$

which shows that the middle-surface stresses increase as the square of the plate curvature. By the same token, it can be shown that the *bending* stresses increase as the *cube* of the plate curvature.

It may be shown from energy considerations that the above solution is valid until β reaches the critical value given by (7.123), at which point (7.125) shows that

$$\begin{aligned} \hat{\kappa}_T &= \hat{\kappa}_T^*, \quad \text{say,}\\ &= 2(1-\nu)^{1/2}/(1+\nu). \end{aligned} \tag{7.126}$$

For values of $|\hat{\kappa}_T|$ greater than $\hat{\kappa}_T^*$, equations (7.121) and (7.123) show that the vanishing of A and C is assured if

$$(1+\nu)\hat{\kappa}_T - (\hat{\kappa}_x + \hat{\kappa}_y) = 0, \tag{7.127}$$

while B is zero whatever the value of $\hat{\kappa}_{xy}$. Thus if we express $\hat{\kappa}_x$, $\hat{\kappa}_y$, $\hat{\kappa}_{xy}$ in terms of principal curvatures $\hat{\kappa}_1, \hat{\kappa}_2$ and their associated angle Ω, say, we see that (7.120) may be written simply as

$$\beta + \hat{\kappa}_1\hat{\kappa}_2 = 0,$$

and (7.127) becomes

$$(1+\nu)\hat{\kappa}_T - (\hat{\kappa}_1 + \hat{\kappa}_2) = 0, \qquad (7.128)$$

while the arbitrary nature of $\hat{\kappa}_{xy}$ means that Ω is arbitrary. Equations (7.128) yield

$$\left.\begin{aligned} \hat{\kappa}_1, \text{say}, &= \tfrac{1}{2}(1+\nu)\{\hat{\kappa}_T + (\hat{\kappa}_T^2 - \hat{\kappa}_T^{*2})^{1/2}\}, \\ \hat{\kappa}_2 &= \tfrac{1}{2}(1+\nu)\{\hat{\kappa}_T - (\hat{\kappa}_T^2 - \hat{\kappa}_T^{*2})^{1/2}\}, \end{aligned}\right\} \tag{7.129}$$

and it is seen that for large values of $|\hat{\kappa}_T|$ one principal curvature tends to $(1+\nu)\hat{\kappa}_T$ while the other tends to zero so that the plate approximates to a developable surface.

7.7 Elliptical plate subjected to certain normal loadings

A comparison of terms in (7.1) shows that the above solution for the deflexion w and force function Φ also applies to an *unheated* plate in which the term κ_T now specifies the magnitude of a *normal load distribution q* given formally by

$$q = -\kappa_T(1+\nu)\nabla^2 D. \tag{7.130}$$

Now

$$\nabla^2 D = -\frac{6D_0}{ab}(\xi + \xi^{-1})F(x, y),$$

where

$$\xi = a/b,$$

$$F(x, y) = \left(1 - \frac{x^2}{a^2} - \frac{y^2}{b^2}\right)\left\{1 - \left(\frac{\xi^2+5}{\xi^2+1}\right)\frac{x^2}{a^2} - \left(\frac{5\xi^2+1}{\xi^2+1}\right)\frac{y^2}{b^2}\right\}. \tag{7.131}$$

Thus if the (unheated) plate is subjected to a (self-equilibrating) load distribution of the form

$$q = q_0 F(x, y), \tag{7.132}$$

the deflexion of the plate is given by the solution in Section 7.6 in which the symbol $\hat{\kappa}_T$ is replaced by $q_0 k a^2 b^2/\{6D_0(1+\nu)(a^2+b^2)\}$.

More generally, it follows from Section 7.6.1 that the solution of (7.1), for the plate defined by (7.111), is given by expressions (7.114), (7.115) whenever the loading can be expressed as a linear combination of $\partial^2 D/\partial x^2$, $\partial^2 D/\partial x\partial y$ and $\partial^2 D/\partial y^2$. Thus if

$$q=\left(1-\frac{x^2}{a^2}-\frac{y^2}{b^2}\right)\left\{q_1\left(1-\frac{5x^2}{a^2}-\frac{y^2}{b^2}\right)+q_2\frac{xy}{ab}+q_3\left(1-\frac{x^2}{a^2}-\frac{5y^2}{b^2}\right)\right\}, \tag{7.133}$$

each of whose components is self-equilibrating, it may be shown that (7.1) may be expressed in the form

$$\left.\begin{aligned}&\Diamond^4(D,\tfrac{1}{2}A'x^2+B'xy+\tfrac{1}{2}C'y^2)=0,\\ &\text{where}\\ &A'=(\beta-\nu)\hat{\kappa}_x-\hat{\kappa}_y+\xi^{-1}\hat{q}_3,\\ &B'=(1-\nu+\beta)\hat{\kappa}_{xy}+\tfrac{1}{8}\hat{q}_2,\\ &C'=(\beta-\nu)\hat{\kappa}_y-\hat{\kappa}_x+\xi\hat{q}_1,\\ &\text{and}\\ &\{\hat{q}_1,\hat{q}_2,\hat{q}_3\}=\left(\frac{kab}{6D_0}\right)\{q_1,q_2,q_3\},\end{aligned}\right\} \tag{7.134}$$

and the other terms are as defined in Section 7.6. Equation (7.4) can again be expressed non-dimensionally as (7.120) and it follows that the complete solution is given by (7.120) and the vanishing of A', B', C' in (7.134). Hence

$$\left.\begin{aligned}\hat{\kappa}_x&=\frac{\xi\hat{q}_1+\xi^{-1}(\beta-\nu)\hat{q}_3}{(1-\nu+\beta)(1+\nu-\beta)},\\ \hat{\kappa}_{xy}&=\frac{-\hat{q}_2}{8(1-\nu+\beta)},\\ \hat{\kappa}_y&=\frac{\xi^{-1}\hat{q}_3+\xi(\beta-\nu)\hat{q}_1}{(1-\nu+\beta)(1+\nu-\beta)},\end{aligned}\right\} \tag{7.135}$$

where β is a root of the quintic

$$\begin{aligned}&\beta(1-\nu+\beta)^2(1+\nu-\beta)^2+\{\xi\hat{q}_1+\xi^{-1}(\beta-\nu)\hat{q}_3\}\\ &\quad\times\{\xi^{-1}\hat{q}_3+\xi(\beta-\nu)q_1\}-\tfrac{1}{64}(1+\nu-\beta)^2\hat{q}_2^2=0.\end{aligned} \tag{7.136}$$

Similar closed-form solutions are possible when the plate has constant initial curvatures.

7.8 Governing differential equations for anisotropic plates

The governing small-deflexion equations (1.97) and (1.98) for the general multi-layered anisotropic plate were derived in Section 1.8. The equation of normal equilibrium (1.97) maintains its validity in the large-deflexion régime and is repeated below for convenience:

$$\mathrm{L}_1 w + \mathrm{L}_3 \Phi = q + \Diamond^4(\Phi, w). \tag{7.137}$$

In the large-deflexion régime, the *condition of compatibility* is given by (7.3) and this results in the following modification to the corresponding small-deflexion equation (1.98):

$$\mathrm{L}_2 \Phi - \mathrm{L}_3 w + \tfrac{1}{2}\Diamond^4(w, w) = 0. \tag{7.138}$$

Some simplification of the L-operators in these equations is possible for the cases discussed in Sections 1.8.3, 1.8.4 and 1.8.5.

7.8.1 Zero coupling between N *and* M

For this important class of plates **B** is zero and the operator L_3 vanishes so that the large-deflexion equations become

$$\mathrm{L}_1 w = q + \Diamond^4(\Phi, w), \tag{7.139}$$

and

$$\mathrm{L}_2 \Phi + \tfrac{1}{2}\Diamond^4(w, w) = 0. \tag{7.140}$$

These equations may be expressed in terms of the displacements, as in Section 7.1.2, by the elimination of N_x, N_y, N_{xy} from (7.137) and (1.32) using the relations

$$\mathbf{N} = \mathbf{A}\boldsymbol{\varepsilon}^0,$$

where $\boldsymbol{\varepsilon}^0$ is defined in terms of displacements by (7.2).

7.8.2 Uniformly loaded long strip

We now consider the large-deflexion analysis of the uniformly loaded infinitely long strip with general multi-layered (coupled) anisotropy. The notation is as shown in Fig. 7.2. This problem is essentially one-dimensional, as in Section 7.2, because the strain pattern does not vary along the length of the strip; this means that $\partial v/\partial y$ is constant throughout the plate and $u, \partial v/\partial x$, w, **M** and **N** are independent of y. Indeed, substitution of $\mathbf{N} = \mathbf{N}(x)$ into the equations of planar equilibrium (1.32), shows that N_x and N_{xy}, but not necessarily N_y, are also constant throughout the plate. By the same token, the equation of normal equilibrium (1.113) is simply

$$\frac{\mathrm{d}^2 M_x}{\mathrm{d}x^2} + N_x \frac{\mathrm{d}^2 w}{\mathrm{d}x^2} + q_0 = 0. \tag{7.141}$$

Now **N**, **M** are given by (1.94) where

$$\left.\begin{aligned} &\begin{bmatrix}\boldsymbol{\varepsilon}^0\\ \boldsymbol{\kappa}\end{bmatrix}=\left[\varepsilon_x^0,\frac{\partial v}{\partial y},\frac{\partial v}{\partial x},\frac{-\mathrm{d}^2w}{\mathrm{d}x^2},0,0\right]^{\mathrm{T}}\\ &\text{and}\\ &\varepsilon_x^0=\frac{\mathrm{d}u}{\mathrm{d}x}+\frac{1}{2}\left(\frac{\mathrm{d}w}{\mathrm{d}x}\right)^2.\end{aligned}\right\} \tag{7.142}$$

Hence, in particular,

$$\left.\begin{aligned} N_x&=A_{11}\varepsilon_x^0+A_{12}\frac{\partial v}{\partial y}+A_{16}\frac{\partial v}{\partial x}-B_{11}\frac{\mathrm{d}^2w}{\mathrm{d}x^2},\\ N_{xy}&=A_{16}\varepsilon_x^0+A_{26}\frac{\partial v}{\partial y}+A_{66}\frac{\partial v}{\partial x}-B_{16}\frac{\mathrm{d}^2w}{\mathrm{d}x^2}\end{aligned}\right\} \tag{7.143}$$

and

$$M_x=B_{11}\varepsilon_x^0+B_{12}\frac{\partial v}{\partial y}+B_{16}\frac{\partial v}{\partial x}-D_{11}\frac{\mathrm{d}^2w}{\mathrm{d}x^2}. \tag{7.144}$$

Now because N_x, N_{xy} and $\partial v/\partial y$ are constants, albeit unknown at this stage, it is convenient to regard (7.143) as determining ε_x^0 and $\partial v/\partial x$ in terms of these constants and $\mathrm{d}^2w/\mathrm{d}x^2$. Thus we find

$$\begin{aligned}(A_{11}A_{66}-A_{16}^2)\varepsilon_x^0=A_{66}N_x-A_{16}N_{xy}+(A_{16}A_{26}-A_{12}A_{66})\partial v/\partial y\\ +(A_{66}B_{11}-A_{16}B_{16})\mathrm{d}^2w/\mathrm{d}x^2\end{aligned} \tag{7.145}$$

and

$$\begin{aligned}&(A_{11}A_{66}-A_{16}^2)\,\partial v/\partial x\\ &\quad=A_{11}N_{xy}-A_{16}N_x+(A_{12}A_{16}-A_{11}A_{26})\,\partial v/\partial y\\ &\qquad+(A_{11}B_{16}-A_{16}B_{11})\mathrm{d}^2w/\mathrm{d}x^2.\end{aligned} \tag{7.146}$$

Substitution of these equations into (7.144) and (7.141) shows that the equation of normal equilibrium can be expressed as

$$\left.\begin{aligned}&\lambda\frac{\mathrm{d}^4w}{\mathrm{d}x^4}+N_x\frac{\mathrm{d}^2w}{\mathrm{d}x^2}+q_0=0,\\ &\text{where}\\ &\lambda=D_{11}-\left(\frac{A_{11}B_{16}^2+A_{66}B_{11}^2-2A_{16}B_{11}B_{16}}{A_{11}A_{66}-A_{16}^2}\right).\end{aligned}\right\} \tag{7.147}$$

Thus far the analysis is quite general in that the boundary conditions have yet to be introduced. To demonstrate the subsequent analysis we now consider the simple case of fully clamped boundaries, for which

$$[u,v,w,\mathrm{d}w/\mathrm{d}x]_{x=\pm\frac{1}{2}a}=0. \tag{7.148}$$

Fully clamped boundaries

It may readily be shown that the solution of (7.147), which yields zero values of w and $\mathrm{d}w/\mathrm{d}x$ at $x = \pm\frac{1}{2}a$, is given by

$$\left.\begin{aligned} w &= \frac{q_0}{N_x}\left\{-\tfrac{1}{2}(x^2 - \tfrac{1}{4}a^2) + \frac{a^2}{4\eta\sinh\eta}\left(\cosh\frac{2\eta x}{a} - \cosh\eta\right)\right\}, \\ &\text{where} \\ \eta^2 &= \frac{a^2 N_x}{4\lambda}. \end{aligned}\right\} \quad (7.149)$$

As for the vanishing of v at the boundaries, we note that integration of (7.146) across the width of the strip, yields the relation

$$A_{11}N_{xy} = A_{16}N_x. \quad (7.150)$$

Finally, the magnitude of N_x is determined by the vanishing of u at the boundaries. Thus, substitution of (7.150) into (7.145) and integration across the width of the strip yields

$$aN_x = \tfrac{1}{2}A_{11}\int_{-\frac{1}{2}a}^{\frac{1}{2}a}\left(\frac{\mathrm{d}w}{\mathrm{d}x}\right)^2 \mathrm{d}x. \quad (7.151)$$

As in Section 7.2, it is more convenient to regard η as the unknown, rather than N_x, so that substitution of (7.149) into (7.151) yields

$$\frac{256\lambda^3}{q_0^2 a^8 A_{11}} = \frac{1}{\eta^8}(1 + \tfrac{5}{12}\eta^2 - \tfrac{3}{4}\eta\coth\eta - \tfrac{1}{4}\eta^2\coth^2\eta), \quad (7.152)$$

and the problem is formally solved.

Other boundary conditions

If the strip forms part of an infinitely wide and uniformly loaded plate supported on a grid of equally spaced y-wise stiffeners that are themselves supported on a grid of x-wise stiffeners, the middle-surface forces in the plate are equilibrated by equal and opposite forces in the supporting structure. The in-plane boundary conditions are now

$$\left.\begin{aligned} &N_x + \frac{k_x}{a}\int_{-\frac{1}{2}a}^{\frac{1}{2}a}\frac{\mathrm{d}u}{\mathrm{d}x}\mathrm{d}x = 0, \\ &\frac{1}{a}\int_{-\frac{1}{2}a}^{\frac{1}{2}a} N_y \mathrm{d}x + k_y\frac{\partial v}{\partial y} = 0, \\ &N_{xy} + \frac{k_{xy}}{a}\int_{-\frac{1}{2}a}^{\frac{1}{2}a}\frac{\partial v}{\partial x}\mathrm{d}x = 0, \end{aligned}\right\} \quad (7.153)$$

where k_x, k_y, k_{xy} are measures of the direct and shear stiffnesses of the supporting structure. Because of symmetry each strip is effectively clamped at its edges so that w is again given by (7.149), but N_x is now determined by (7.153).

References

Kármán, Th. von. *Enzyklopädie der mathematischen Wissenschaften.* Vol. 4. 1910.

Mansfield, E. H. Large-deflexion torsion and flexure of initially curved strips. *Proc. R. Soc. Lond.*, A. **334**, pp. 279–98 (1973).

Way, S. Bending of circular plates with large deflection. A.S.M.E. Trans., A.P.M.-56-12, **56**(8), pp. 627–36 (August 1934).

Additional references

Ashwell, D. G. The boundary layer in the bending of inextensible plates and shells, *Quart. J. Mech. Appl. Math.*, **16**, pp. 179–91 (1963).

Cox, H. L. The buckling of a flat rectangular plate under axial compression and its behaviour after buckling. Part I and II. *Aero. Res. Council R. & M. Nos. 2041*, 2175, H.M.S.O. (1945).

Fung, Y. C., and Wittrick, W. H. The anti-clastic curvature of a strip with lateral thickness variation. *J. Appl. Mech.*, **21**, pp. 351–8 (December 1954).

______. A boundary layer phenomenon in the large deflexion of thin plates. *Quart J. Mech. Appl. Math.*, **8**, pp. 191–210 (1955).

Levy, S. Bending of rectangular plates with large deflections. *N.A.C.A. Rep.* No. 737 (1942).

______. Square plate with clamped edges under normal pressure producing large deflections. *N.A.C.A. Rep.* No. 740 (1942).

Mansfield, E. H. The large-deflexion behaviour of a thin strip of lenticular section. *Quart. J. Mech. Appl. Maths.*, **12**(4), pp. 421–30 (November 1959).

______. Bending, buckling and curling of a heated elliptical plate. *Proc. R. Soc. Lond.* Ser. A. **288**, pp. 396–417 (1965).

Reissner, E. Finite twisting and bending of thin rectangular elastic plates. *J. Appl. Mech.*, **24**, pp. 391–6 (September 1957).

8

Approximate methods in large-deflexion analysis

Here we discuss some approximate methods of analysis of the large-deflexion behaviour of plates of constant thickness, focusing attention first on the case of isotropy. The loading considered is either a uniformly distributed normal load, or a compressive or shear load in the plane of the plate in excess of that necessary to cause buckling.

In Sections 8.1 and 8.2 we start from the displacement equations (7.17) and (7.18) whose approximate solution we obtain by a perturbation technique. In Section 8.3 we see how approximate solutions may be obtained from the principle of minimum potential energy. The anisotropic plate is discussed in Section 8.4.

8.1 Perturbation method for normally loaded plates

In this method, a solution of (7.17) and (7.18) is sought in the form of expansions in ascending powers of a convenient deflexion Δ; in a plate with two-fold symmetry it would be convenient to let Δ be the central deflexion. It is then assumed that the quantities q_0, w, u, v can be expressed in the form

$$\left.\begin{aligned} q_0 &= \alpha_1\Delta + \alpha_3\Delta^3 + \cdots \\ w &= w_1(x,y)\Delta + w_3(x,y)\Delta^3 + \cdots \end{aligned}\right\} \quad (8.1)$$

$$\left.\begin{aligned} u &= u_2(x,y)\Delta^2 + u_4(x,y)\Delta^4 + \cdots \\ v &= v_1(x,y)\Delta^2 + v_4(x,y)\Delta^4 + \cdots \end{aligned}\right\} \quad (8.2)$$

where q_0 is the intensity of loading at the point (x_0, y_0), say, whose deflexion is Δ, the α_n are constants, and w_n, u_n, v_n are functions of x, y to be determined. Only odd powers of Δ are required in (8.1) because a change in sign of q produces a change in sign of w; by the same token only even powers of Δ occur in (8.2) because a change in sign of q, and hence of Δ, does not affect the displacements u, v. Further, in virtue of the definition of Δ it is also necessary that

$$\left.\begin{aligned} &w_1(x_0, y_0) = 1, \\ &w_3(x_0, y_0) = w_5(x_0, y_0) = \cdots = 0. \end{aligned}\right\} \quad (8.3)$$

Also, each of the functions w_n, u_n, v_n must satisfy the boundary conditions.

Substitution of (8.1) and (8.2) into (7.17) and equating terms of order Δ results in the small-deflexion equation

$$D\nabla^4 w - q = 0 \tag{8.4}$$

whose solution, assumed known, can be expressed in the form

$$w = w_1(x, y)\Delta. \tag{8.5}$$

Next, substitution of (8.1) and (8.2) into (7.18) and equating terms of order Δ^2 yields the following linear equations for determining u_2, v_2:

$$\left.\begin{aligned}
&\frac{\partial^2 u_2}{\partial x^2}+\left(\frac{1-v}{2}\right)\frac{\partial^2 u_2}{\partial y^2}+\left(\frac{1+v}{2}\right)\frac{\partial^2 v_2}{\partial x\partial y}\\
&\qquad+\frac{\partial w_1}{\partial x}\frac{\partial^2 w_1}{\partial x^2}+\left(\frac{1-v}{2}\right)\frac{\partial w_1}{\partial x}\frac{\partial^2 w_1}{\partial y^2}+\left(\frac{1+v}{2}\right)\frac{\partial w_1}{\partial y}\frac{\partial^2 w_1}{\partial x\partial y}=0\\
&\frac{\partial^2 v_2}{\partial y^2}+\left(\frac{1-v}{2}\right)\frac{\partial^2 v_2}{\partial x^2}+\left(\frac{1+v}{2}\right)\frac{\partial^2 u_2}{\partial x\partial y}\\
&\qquad+\frac{\partial w_1}{\partial y}\frac{\partial^2 w_1}{\partial y^2}+\left(\frac{1-v}{2}\right)\frac{\partial w_1}{\partial y}\frac{\partial^2 w_1}{\partial x^2}+\left(\frac{1+v}{2}\right)\frac{\partial w_1}{\partial x}\frac{\partial^2 w_1}{\partial x\partial y}=0.
\end{aligned}\right\} \tag{8.6}$$

Substitution of (8.1) and (8.2) in (7.17) and equating terms of order Δ^3 now yields the followng linear equation for determining w_3:

$$\begin{aligned}
\frac{t^2}{12}\nabla^4 w_3 = &\frac{(1-v^2)\alpha_3}{Et}+\frac{\partial^2 w_1}{\partial x^2}\left\{\frac{\partial u_2}{\partial x}+v\frac{\partial v_2}{\partial y}+\frac{1}{2}\left(\frac{\partial w_1}{\partial x}\right)^2+\tfrac{1}{2}v\left(\frac{\partial w_1}{\partial y}\right)^2\right\}\\
&+\frac{\partial^2 w_1}{\partial y^2}\left\{\frac{\partial v_2}{\partial y}+v\frac{\partial u_2}{\partial x}+\frac{1}{2}\left(\frac{\partial w_1}{\partial y}\right)^2+\tfrac{1}{2}v\left(\frac{\partial w_1}{\partial x}\right)^2\right\}\\
&+(1-v)\frac{\partial^2 w_1}{\partial x\partial y}\left(\frac{\partial u_2}{\partial y}+\frac{\partial v_2}{\partial x}+\frac{\partial w_1}{\partial x}\frac{\partial w_1}{\partial y}\right).
\end{aligned} \tag{8.7}$$

By the same token, substitution of (8.1) and (8.2) into (7.18) and equating terms of order Δ^4 yields two linear equations for determining u_4, v_4 and the cycle of operations may be repeated. For plates whose boundary supports resist movement in the plane of the plate, the first two terms in the series for w suffice to determine the deflexion well into the large-deflexion régime. The reason for such agreement stems from the fact that for such plates, as is shown in Section 9.1, the deflexion increases asymptotically as $q^{1/3}$ for large values of q, and this is in accord with an equation of the form $q = \alpha_1\Delta + \alpha_3\Delta^3$.

8.1.1 *Uniformly loaded clamped elliptical plate*

The perturbation method was employed by Chien (1947) in discussing the uniformly loaded clamped circular plate. Here, as an illustrative example, we outline the treatment by Nash and Cooley (1959) of the uniformly loaded clamped elliptical plate. The small-deflexion solution for such a plate is given by (3.48) and therefore

$$w_1(x, y) = \left(1 - \frac{x^2}{a^2} - \frac{y^2}{b^2}\right)^2$$

and

$$\alpha_1 = 8D\left(\frac{3}{a^4} + \frac{3}{b^4} + \frac{2}{a^2 b^2}\right). \tag{8.8}$$

Substitution of (8.8) into (8.6) yields two simultaneous linear equations for determining u_2, v_2 which must satisfy the following boundary condition

$$u_2 = v_2 = 0 \quad \text{along} \; \frac{x^2}{a^2} + \frac{y^2}{b^2} = 1. \tag{8.9}$$

A suitable form for the displacements which satisfies (8.9) is given by

$$\left.\begin{aligned} u_2 &= x\left(1 - \frac{x^2}{a^2} - \frac{y^2}{b^2}\right)(A_1 + A_2 x^2 + A_3 y^2 + A_4 x^4 + A_5 y^4 + A_6 x^2 y^2) \\ v_2 &= y\left(1 - \frac{x^2}{a^2} - \frac{y^2}{b^2}\right)(B_1 + B_2 x^2 + B_3 y^2 + B_4 x^4 + B_5 y^4 + B_6 x^2 y^2) \end{aligned}\right\} \tag{8.10}$$

and the coefficients $A_1, \ldots, A_6$, $B_1, \ldots, B_6$ are determined by substituting (8.10) into (8.6) and equating powers of x and y. At this stage it is preferable to introduce the numerical value of a/b. Next, substitution of (8.8) and (8.10) into (8.7) yields an equation for determining w_3. The solution of this equation may be sought in the form

$$w_3 = \left(1 - \frac{x^2}{a^2} - \frac{y^2}{b^2}\right)^2 (C_1 x^2 + C_2 y^2 + C_3 x^4 + C_4 y^4 + C_5 x^2 y^2) \tag{8.11}$$

which satisfies (8.5) together with the boundary conditions

$$w_3 = \frac{\partial w_3}{\partial x} = \frac{\partial w_3}{\partial y} = 0 \quad \text{along} \; \frac{x^2}{a^2} + \frac{y^2}{b^2} = 1.$$

The coefficients α_3 and $C_1, \ldots, C_5$ are determined by equating like powers of x and y.

8.2 Perturbation method in post-buckling problems

Stein (1959) has shown that a technique, similar to that discussed in Section 8.1, may be used to investigate the post-buckling behaviour of plates, and he has applied the method to simply supported rectangular plates subjected to various combinations of compressive forces in the plane of the plate. In this technique it is necessary to expand the displacements u, v, w about the point of buckling in powers of a suitable parameter. Stein points out that there is some freedom in the choice of this parameter, and for the uniaxial compression problem chooses the parameter $\delta = \{(P - P_{cr})/P_{cr}\}^{1/2}$. This form is suitable, as it is known that immediately after buckling the deflexion increases in proportion to δ. Here, however, we follow the notation of Section 8.1 and expand the displacements in powers of the deflexion Δ at a chosen point (x_0, y_0), and write

$$\left.\begin{aligned} w &= w_1(x,y)\Delta + w_3(x,y)\Delta^3 + \cdots \\ u &= u_0(x,y) + u_2(x,y)\Delta^2 + \cdots \\ v &= v_0(x,y) + v_2(x,y)\Delta^2 + \cdots. \end{aligned}\right\} \tag{8.12}$$

Only odd powers of w and even powers of u, v are required, because a change in the sign of w does not affect the displacements in the plane of the plate. The terms u_0, v_0 are the displacements in the plane of the plate at the onset of buckling. As in Section 8.1 the functions w_n must satisfy (8.3) and the term w_n, u_n, v_n must each satisfy the boundary conditions.

Substitution of (8.12) into (7.18) and equating terms independent of Δ yields the following relations for determining the form, but not the magnitudes, of u_0, v_0

$$\left.\begin{aligned} \frac{\partial}{\partial x}\left(\frac{\partial u_0}{\partial x} + \frac{\partial v_0}{\partial y}\right) + \left(\frac{1-\nu}{1-\nu}\right)\nabla^2 u_0 = 0 \\ \frac{\partial}{\partial y}\left(\frac{\partial u_0}{\partial x} + \frac{\partial v_0}{\partial y}\right) + \left(\frac{1-\nu}{1+\nu}\right)\nabla^2 v_0 = 0. \end{aligned}\right\} \tag{8.13}$$

These are simply the equations of plane stress expressed in terms of displacements.

Next, substitution of (8.12) into (7.17) and equating terms of order Δ yields the small-deflexion equation

$$\begin{aligned} \frac{t^2}{12}\nabla^4 w_1 = {} & \frac{\partial^2 w_1}{\partial x^2}\left(\frac{\partial u_0}{\partial x} + \nu\frac{\partial v_0}{\partial y}\right) + \frac{\partial^2 w_1}{\partial y^2}\left(\frac{\partial v_0}{\partial y} + \nu\frac{\partial u_0}{\partial x}\right) \\ & + (1-\nu)\frac{\partial^2 w_1}{\partial x \partial y}\left(\frac{\partial u_0}{\partial y} + \frac{\partial v_0}{\partial x}\right) \end{aligned} \tag{8.14}$$

whose solution determines the magnitude of u_0, v_0 and the function w_1.

Substitution of (8.12) into (7.18) and equating terms of order Δ^2 now yields equation (8.6), from which the functions u_2, v_2 are to be determined – apart from an arbitrary term.

Similarly, equating terms of order Δ^3 yields the following equation:

$$\begin{aligned}
&\frac{t^2}{12}\nabla^4 w_3 - \frac{\partial^2 w_3}{\partial x^2}\left(\frac{\partial u_0}{\partial x} + \nu\frac{\partial v_0}{\partial y}\right) - \frac{\partial^2 w_3}{\partial y^2}\left(\frac{\partial v_0}{\partial y} + \nu\frac{\partial u_0}{\partial x}\right) \\
&\quad - (1-\nu)\frac{\partial^2 w_3}{\partial x \partial y}\left(\frac{\partial u_0}{\partial y} + \frac{\partial v_0}{\partial x}\right) \\
&= \frac{\partial^2 w_1}{\partial x^2}\left\{\frac{\partial u_2}{\partial x} + \nu\frac{\partial v_2}{\partial y} + \frac{1}{2}\left(\frac{\partial w_1}{\partial x}\right)^2 + \tfrac{1}{2}\nu\left(\frac{\partial w_1}{\partial y}\right)^2\right\} \\
&\quad + \frac{\partial^2 w_1}{\partial y^2}\left\{\frac{\partial v_2}{\partial y} + \nu\frac{\partial u_2}{\partial x} + \frac{1}{2}\left(\frac{\partial w_1}{\partial y}\right)^2 + \tfrac{1}{2}\nu\left(\frac{\partial w_1}{\partial x}\right)^2\right\} \\
&\quad + (1-\nu)\frac{\partial^2 w_1}{\partial x \partial y}\left(\frac{\partial u_2}{\partial y} + \frac{\partial v_2}{\partial x} + \frac{\partial w_1}{\partial x}\frac{\partial w_1}{\partial y}\right)
\end{aligned} \tag{8.15}$$

from which w_3 is determined together with the previously arbitrary terms in u_2, v_2. The cycle of such operations may be repeated to obtain ever increasing accuracy. At this stage, however, it is more useful to demonstrate the method in detail for a simple example.

8.2.1 *Post-buckling behaviour of compressed square plate*

To demonstrate the method, we now consider the post-buckling behaviour of a square plate simply supported along the edges and subjected to a load causing one pair of opposite edges to approach each other by a fixed amount while the distance between the other pair of edges remains constant. We further stipulate that all edges are constrained to remain straight, and that there is zero edge shear stress in the plane of the plate. If the plate is bounded by the lines $x=0, a$ and $y=0, a$ the boundary conditions are then given by

$$\left.\begin{aligned}
&\frac{\partial u}{\partial y} = \frac{\partial v}{\partial x} = w = \frac{\partial^2 w}{\partial x^2} = 0, \quad \text{along } x = 0, a \\
&\frac{\partial u}{\partial x} = v = w = \frac{\partial^2 w}{\partial y^2} = 0, \quad \text{along } y = 0, a.
\end{aligned}\right\} \tag{8.16}$$

The centre of the plate $(\frac{1}{2}a, \frac{1}{2}a)$ is defined as the point (x_0, y_0) at which the deflexion is Δ. Attention is also confined to plates exhibiting a single buckle, although for high values of the compression two or more buckles occur in the direction of the compression. There is, however, no difficulty in extending the analysis to include more buckles nor in considering the general case of a rectangular plate. It must be admitted, however, that the particular boundary conditions considered here are specially chosen to yield simple results.

The solution of (8.13) which satisfies (8.16) may be written down by inspection:

$$\left.\begin{aligned} v_0 &= 0 \\ u_0 &= -k_0 x, \end{aligned}\right\} \tag{8.17}$$

but it is to be noted that the constant k_0 is at present unknown. This is because the terms u_0, v_0 apply only to conditions at the onset of buckling which have yet to be determined. Substitution of (8.17) into (8.14) now yields the small-deflexion equation:

$$\frac{t^2}{12}\nabla^2 w_1 = -k_0\left(\frac{\partial^2 w_1}{\partial x^2} + \nu\frac{\partial^2 w_1}{\partial y^2}\right) \tag{8.18}$$

whose solution (apart from the trivial case with w_1 zero) determines k_0 and is given by

$$\left.\begin{aligned} w_1 &= \sin\frac{\pi x}{a}\sin\frac{\pi y}{a}, \\ k_0 &= \frac{\pi^2 t^2}{3(1+\nu)a^2}. \end{aligned}\right\} \tag{8.19}$$

Thus far the analysis is identical with small-deflexion theory and the magnitude of the central deflexion Δ is still arbitrary. Substituting (8.19) into (8.6) and simplifying yields

$$\left.\begin{aligned} &\frac{\partial^2 u_2}{\partial x^2} + \left(\frac{1-\nu}{2}\right)\frac{\partial^2 u_2}{\partial y^2} + \left(\frac{1+\nu}{2}\right)\frac{\partial^2 v_2}{\partial x \partial y} \\ &\qquad -\frac{1}{4}\left(\frac{\pi}{a}\right)^3\left\{1-\nu-2\cos\frac{2\pi y}{a}\right\}\sin\frac{2\pi x}{a} = 0 \\ &\frac{\partial^2 v_2}{\partial y^2} + \left(\frac{1-\nu}{2}\right)\frac{\partial^2 v_2}{\partial x^2} + \left(\frac{1+\nu}{2}\right)\frac{\partial^2 u_2}{\partial x \partial y} \\ &\qquad -\frac{1}{4}\left(\frac{\pi}{a}\right)^3\left\{1-\nu-2\cos\frac{2\pi x}{a}\right\}\sin\frac{2\pi y}{a} = 0 \end{aligned}\right\} \tag{8.20}$$

whose solution is given by

$$u_2 = -k_2 x - \frac{\pi}{16a}\left(1-\nu-\cos\frac{2\pi y}{a}\right)\sin\frac{2\pi x}{a} \tag{8.21}$$

$$v_2 = -\frac{\pi}{16a}\left(1 - \nu - \cos\frac{2\pi x}{a}\right)\sin\frac{2\pi y}{a},$$

where the constant k_2 is at present unknown.

Substitution of (8.17), (8.19) and (8.21) in (8.15) yields the following linear differential equation whose solution determines k_2 and w_3:

$$\frac{t^2}{12}\nabla^4 w_3 + \frac{\pi^2 t^2}{3(1+\nu)a^2}\left(\frac{\partial^2 w_3}{\partial x^2} + \nu\frac{\partial^2 w_3}{\partial y^2}\right)$$
$$= \frac{\pi^2(1+\nu)}{a^2}\left\{k_2 - \frac{\pi^2}{8a^2}(3-\nu)\right\}\sin\frac{\pi x}{a}\sin\frac{\pi y}{a}$$
$$+ \frac{\pi^4(1-\nu^2)}{16a^4}\left(\sin\frac{\pi x}{a}\sin\frac{3\pi y}{a} + \sin\frac{3\pi x}{a}\sin\frac{\pi y}{a}\right). \tag{8.22}$$

It is to be noted that the term $\sin \pi x/a \sin \pi y/a$ is a complementary solution of (8.22), but it also occurs on the right-hand side of (8.22). This may be shown to lead to a mathematical impasse and we can infer that the coefficient of $\sin \pi x/a \sin \pi y/a$ in (8.22) must be zero. Thus

$$k_2 = \frac{\pi^2}{8a^2}(3-\nu). \tag{8.23}$$

It may also be verified that the solution of (8.22) which satisfies (8.5) and the boundary conditions (8.16) is now given by

$$w_3 = (A+B)\sin\frac{\pi x}{a}\sin\frac{\pi y}{a} + A\sin\frac{\pi x}{a}\sin\frac{3\pi y}{a} + B\sin\frac{3\pi x}{a}\sin\frac{\pi y}{a}$$

where

$$A = \frac{3(1-\nu)(1+\nu)^2}{16(24+25\nu-9\nu^2)t^2},$$

$$B = \frac{3(1-\nu)(1+\nu)^2}{16(16+25\nu-\nu^2)t^2}. \tag{8.24}$$

The cycle of operations for determining u_4, v_4 and w_5 proceeds on lines similar to those used to determine u_2, v_2 and w_3, and is not given here. The solution obtained so far is accurate in w to terms of order Δ^3 and accurate in u, v to terms of order Δ^2. Thus it is possible, for example, to relate the amount by which the loaded edges approach each other, δu, say, to the magnitude of the central deflexion Δ:

$$\delta u - \delta u_{cr} = ak_2\Delta^2 + O(\Delta^4)$$
$$= \frac{\pi^2(3-\nu)\Delta^2}{8a} + O(\Delta^4). \tag{8.25}$$

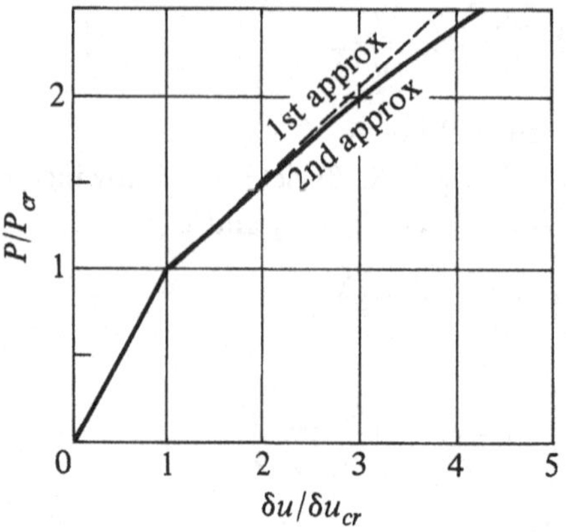

Fig. 8.1

However, what is of more practical importance is the relationship between $(\delta u - \delta u_{cr})$ and $(P - P_{cr})$ where P is the total applied compressive load. This is found readily from (7.2), which gives

$$P = -\frac{Et}{1-\nu^2}\int_0^a \left\{\frac{\partial u}{\partial x} + \nu\frac{\partial v}{\partial y} + \frac{1}{2}\left(\frac{\partial w}{\partial x}\right)^2 + \tfrac{1}{2}\nu\left(\frac{\partial w}{\partial y}\right)^2\right\} \mathrm{d}y \tag{8.26}$$

so that, from (8.12), (8.17) and (8.19) we find

$$P - P_{cr} = -\frac{Et\Delta^2}{1-\nu^2}\int_0^a \left\{\frac{\partial u_2}{\partial x} + \nu\frac{\partial v_2}{\partial y} + \frac{1}{2}\left(\frac{\partial w_1}{\partial x}\right)^2 + \tfrac{1}{2}\nu\left(\frac{\partial w_1}{\partial y}\right)^2\right\} \mathrm{d}y + \mathrm{O}(\Delta^4) = \frac{\pi^2 Et\Delta^2}{4(1+\nu)a} + \mathrm{O}(\Delta^4). \tag{8.27}$$

Equations (8.25) and (8.27) may be combined to give

$$\frac{P - P_{cr}}{\delta u - \delta u_{cr}} = \frac{2Et}{(1+\nu)(3-\nu)} + \mathrm{O}(\Delta^2). \tag{8.28}$$

Now $(P - P_{cr})/(\delta u - \delta u_{cr})$ is the direct stiffness of the plate immediately after the onset of buckling and may be compared with the value $Et/(1-\nu^2)$ prior to buckling. Thus for the particular example considered here:

$$\frac{\text{stiffness of plate immediately after buckling}}{\text{stiffness of plate prior to buckling}} = \frac{2(1-\nu)}{3-\nu}. \tag{8.29}$$

The stiffness after buckling of all flat plates decreases slightly as P/P_{cr} increases. However, as indicated in Fig. 8.1, if the wavelength of the buckles does not change, the reduction is small in the range $1 < P/P_{cr} < 2$ and a good estimate can be expected from an analysis which relates the load P to the deflexion Δ with an error of $\mathrm{O}(\Delta^6)$. If the wavelength of the buckles changes, there may be a marked drop in the 'tangent' stiffness, but this too may be analysed by the perturbation method.

8.3 An energy method

The principle of minimum potential energy may also be used in the large deflexion analysis of plates. The strain energy of bending is given by (6.4) and the strain energy due to stretching of the middle surface is

$$\frac{E}{2(1-\nu^2)}\iint t\{\varepsilon_x^2+\varepsilon_y^2+2\nu r\varepsilon_x\varepsilon_y+\tfrac{1}{2}(1-\nu)\varepsilon_{xy}^2\}\,\mathrm{d}x\,\mathrm{d}y$$

where $\varepsilon_x, \varepsilon_y, \varepsilon_{xy}$ are given by (7.2).

The total potential energy may therefore be expressed in terms of the displacements u, v, w as follows:

$$\begin{aligned}\Pi=\frac{1}{2}&\iint D\{(\nabla^2 w)^2-(1-\nu)\,\Diamond^4(w,w)\}\,\mathrm{d}x\,\mathrm{d}y\\&+\frac{E}{2(1-\nu^2)}\iint t\left[\left(\frac{\partial u}{\partial x}\right)^2+\frac{\partial u}{\partial x}\left(\frac{\partial w}{\partial x}\right)^2+\left(\frac{\partial v}{\partial y}\right)^2+\frac{\partial v}{\partial y}\left(\frac{\partial w}{\partial y}\right)^2\right.\\&+\frac{1}{4}\left\{\left(\frac{\partial w}{\partial x}\right)^2+\left(\frac{\partial w}{\partial y}\right)^2\right\}^2+2\nu\left\{\frac{\partial u}{\partial x}\frac{\partial v}{\partial y}+\frac{1}{2}\frac{\partial v}{\partial y}\left(\frac{\partial w}{\partial x}\right)^2\right.\\&\left.+\frac{1}{2}\frac{\partial u}{\partial x}\left(\frac{\partial w}{\partial y}\right)^2\right\}+\tfrac{1}{2}(1-\nu)\left\{\left(\frac{\partial u}{\partial y}\right)^2+2\frac{\partial u}{\partial y}\frac{\partial v}{\partial x}+\left(\frac{\partial v}{\partial x}\right)^2\right.\\&\left.\left.+2\left(\frac{\partial u}{\partial y}+\frac{\partial v}{\partial x}\right)\frac{\partial w}{\partial x}\frac{\partial w}{\partial y}\right\}\right]\mathrm{d}x\,\mathrm{d}y-\iint qw\,\mathrm{d}x\,\mathrm{d}y+\Pi_e+U_e.\end{aligned}\tag{8.30}$$

where Π_e and U_e are as defined by (6.12), (6.13).

The energy method was used by Cox (1933) in discussing the post-buckling behaviour of a rectangular strip under end load. The importance of this analysis lay in the choice of the deflexion w. For example, when the edges were simply supported, it was assumed that the deflexion, which is symmetrical about the centre line $y=\frac{1}{2}b$, is given by

$$w=w_0\sin\frac{n\pi x}{a}\sin\frac{\pi y}{\alpha b}\tag{8.31a}$$

in the range $0\leqslant y\leqslant\frac{1}{2}\alpha b$, where α depends on the applied loading, and

$$w=w_0\sin\frac{n\pi x}{a}\tag{8.31b}$$

in the range $\frac{1}{2}\alpha b<y\leqslant\frac{1}{2}b$.

In the immediate post-buckling phase, the coefficient α is unity, but as the loading increases the deflected mode changes so that over the central region of the plate the mode is a developable surface. If allowance had not been made to include such a developable zone, the resulting analysis would have predicted an overall stiffness under a high loading in excess of the true value. This is because the strain energy due to the middle-surface forces would have been overestimated.

In practice, the deflected form does not contain an exactly developable zone, nor, of course, does it contain a discontinuity in $\partial^2 w/\partial y^2$ as does (8.31) at the line $y = \frac{1}{2}\alpha b$. Nevertheless, (8.31) is probably the best available representation if the limit of disposable parameters (w_0, n, α) is three. To overcome the objection of the discontinuous character of $\partial^2 w/\partial y^2$ Koiter (1943) suggested the form

$$w = w_0 \sin\frac{n\pi x}{a}\left(\frac{2y}{\alpha b} + \frac{1}{\pi}\sin\frac{2\pi y}{\alpha b}\right),$$

valid in the range $0 \leqslant y \leqslant \frac{1}{2}\alpha b$, but the resulting improvement in accuracy was negligible.

8.3.1 Post-buckling behaviour of long strip under compression and shear

The post-buckling behaviour of a long simply supported strip under combined shear and longitudinal and lateral compression or tension has been investigated by Koiter (1944), Van der Neut and Floor (1948), and Floor and Burgerhout (1951). The presence of shear necessitates a slight modification to (8.31) and the deflected form (Fig. 8.2) is represented by

$$\left.\begin{array}{l} w = w_0 \sin\dfrac{\pi(x - my)}{L}\sin\dfrac{\pi y}{\alpha b} \\ \text{in the range } 0 \leqslant y \leqslant \frac{1}{2}\alpha b, \text{ and} \\ w = w_0 \sin\dfrac{\pi(x - my)}{L} \end{array}\right\} \qquad (8.32)$$

in the range $\frac{1}{2}\alpha b < y \leqslant \frac{1}{2}b$.

Here L is the longitudinal half-wave length and the parameter m determines the orientation of the nodal lines. In practice the nodal lines are not straight, except in the absence of shear, but they become progressively straighter as the loading is increased; furthermore (8.32) does not satisfy the edge conditions of simple support, except when m is zero. As a consequence, the greatest errors arising from the use of (8.32) occur at the onset and shortly after buckling when the loading is predominantly shear. A comparison with the known exact solution for buckling under

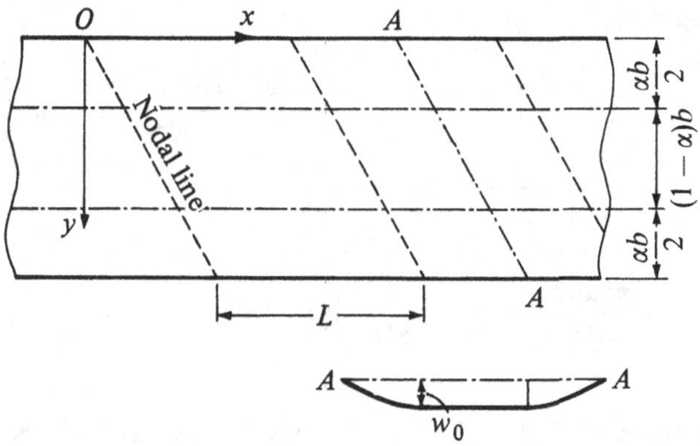

Fig. 8.2

pure shear, due to Southwell and Skan (1924), then shows that the maximum error does not exceed 6 per cent. In many instances, however, we are concerned with the post-buckling behaviour of a *continuous* plate simply supported at a series of lines ($y = 0, b, 2b$, etc.). This condition should strictly be described as one of *continuity*, and it is a more constrained condition than that of simple support. It follows that for continuous plates, (8.32) is in error at most by about 3 per cent.

In addition to choosing a suitable form for the deflexion, it is necessary to prescribe forms for the displacements u, v. The following forms were derived by Koiter and were chosen to minimize the elastic energy and to satisfy the condition that the edges of the strip remain straight:

In the range $0 \leqslant y \leqslant \frac{1}{2}\alpha b$,

$$\left.\begin{aligned} u = &-\varepsilon_1 x + \gamma y - \frac{\pi w_0^2}{16L}\left(1 - \cos\frac{2\pi y}{\alpha b}\right)\sin\frac{2\pi(x - my)}{L} \\ &- \frac{m\pi^2 w_0^2}{4L^2}(1-\alpha)y - \frac{m\pi\alpha b w_0^2}{8L^2}\sin\frac{2\pi y}{\alpha b}, \\ v = &-\varepsilon_2 y \frac{m\pi w_0^2}{16L}\left(1 - \cos\frac{2\pi y}{\alpha b}\right)\sin\frac{2\pi(x - my)}{L} \\ &+ \frac{\pi w_0^2}{16\alpha b}\sin\frac{2\pi y}{\alpha b}\cos\frac{2\pi(x - my)}{L} \\ &+ \frac{\pi^2 w_0^2}{8L^2}(1-\alpha)\left(\nu + m^2 - \frac{L^2}{\alpha^2 b^2}\right)y \\ &+ \frac{\pi\alpha b w_0^2}{16L^2}\left(\nu + m^2 - \frac{L^2}{\alpha^2 b^2}\right)\sin\frac{2\pi y}{\alpha b}. \end{aligned}\right\} \quad (8.33a)$$

In the range $\tfrac{1}{2}\alpha b < y \leqslant \tfrac{1}{2}b$,

$$\left.\begin{aligned}
u &= -\varepsilon_1 x + \gamma y - \frac{\pi w_0^2}{8L}\sin\frac{2\pi(x-my)}{L} + \frac{m\pi^2\alpha w_0^2}{4L^2}(y-\tfrac{1}{2}b),\\
v &= -\varepsilon_2 y + \frac{m\pi w_0^2}{8L}\sin\frac{2\pi(x-my)}{L}\\
&\quad -\frac{\pi^2\alpha w_0^2}{8L^2}\left(v + m^2 - \frac{L^2}{\alpha^2 b^2}\right)(y-\tfrac{1}{2}b).
\end{aligned}\right\} \tag{8.33b}$$

In equations (8.33a), (8.33b) the terms ε_1 and ε_2 are the overall compressive strains in the longitudinal (x-) and lateral (y-) directions of the plate, and γ is the overall shear strain of the plate.

Substitution of (8.32), (8.33a) and (8.33b) in (8.30) for the average strain energy U per unit area of plate yields the relation:

$$\begin{aligned}
2EU/t &= \sigma_1^2 - 2v\sigma_1\sigma_2 + \sigma_2^2 + 2(1+v)\tau^2\\
&\quad + E^2F^2\left\{(1-\tfrac{5}{8}\alpha-\tfrac{1}{4}\alpha^2)H^2 + \frac{1}{8\alpha^3(1-v^2)}\right\}\\
&\quad + \frac{\pi^2E^2F^2t^2}{6b^2(1-v^2)}\{(1-\tfrac{1}{2}\alpha)(1+m^2)^2H^2\\
&\quad + (1+3m^2)H/\alpha + 1/\alpha^3\}
\end{aligned} \tag{8.34}$$

where

$$F = \frac{\pi^2 w^2}{4L^2},$$

$$H = b^2L^2,$$

and σ_1, σ_2 are the average tensile stresses in the longitudinal and lateral directions of the plate and are given by

$$\left.\begin{aligned}
(\sigma_1 - v\sigma_2)/E &= -\varepsilon_1 + (1-\tfrac{1}{2}\alpha)HF,\\
(\sigma_2 - v\sigma_1)/E &= -\varepsilon_2 + (1-\tfrac{1}{2}\alpha)m^2HF + \tfrac{1}{2}F/\alpha,\\
2(1+v)\tau/E &= \gamma - 2(1-\tfrac{1}{2}\alpha)mHF.
\end{aligned}\right\} \tag{8.35}$$

The parameters w_0, L, m, α or, alternatively, F, H, m, α are to be determined from the conditions

$$\frac{\partial U}{\partial F} = \frac{\partial U}{\partial H} = \frac{\partial U}{\partial m} = \frac{\partial U}{\partial \alpha} = 0 \tag{8.36}$$

in which it is to be assumed that $\varepsilon_1, \varepsilon_2, \gamma$ are given constants.

8.4 Anisotropic plates

The perturbation method of analysis may also be used for anisotropic plates. We first express the governing equations in terms of the displacements u, v, w. In this connection we note from (7.2) that the strains in the middle surface are given by

$$\boldsymbol{\varepsilon}^0 = \left[\frac{\partial u}{\partial x} + \frac{1}{2}\left(\frac{\partial w}{\partial x}\right)^2, \frac{\partial v}{\partial y} + \frac{1}{2}\left(\frac{\partial w}{\partial y}\right)^2, \frac{\partial u}{\partial y} + \frac{\partial v}{\partial x} + \frac{\partial w}{\partial x}\frac{\partial w}{\partial y}\right]^{\mathrm{T}}, \tag{8.37}$$

and hence, for the general multi-layered anisotropic plate, (1.94) and (8.37) enable us to express $\mathbf{N}, \mathbf{M}$ in terms of u, v, w. This, in turn, enables us to express the equilibrium conditions, namely (1.32) and (1.113), in terms of u, v, w.

For the general multi-layered plate the presence of the matrix $\mathbf{B}$ in (1.94) adds to the complexity of the perturbation method because both odd and even powers of Δ are needed in the expansions for u and v.

8.4.1 Zero coupling between N *and* M

For this important class of plates, $\mathbf{B}$ is zero and hence the middle-surface forces are given simply by

$$\mathbf{N} = \mathbf{A}\boldsymbol{\varepsilon}^0, \tag{8.38}$$

where $\boldsymbol{\varepsilon}^0$ is given by (8.37). The equations of equilibrium in the plane of the plate may therefore be expressed in terms of the displacements as follows:

$$\left.\begin{aligned}
&\frac{\partial}{\partial x}\left[A_{11}\left\{\frac{\partial u}{\partial x} + \frac{1}{2}\left(\frac{\partial w}{\partial x}\right)^2\right\} + A_{12}\left\{\frac{\partial v}{\partial y} + \frac{1}{2}\left(\frac{\partial w}{\partial y}\right)^2\right\}\right.\\
&\qquad\left. + A_{16}\left\{\frac{\partial u}{\partial y} + \frac{\partial v}{\partial x} + \frac{\partial w}{\partial x}\frac{\partial w}{\partial y}\right\}\right] + \frac{\partial}{\partial y}\left[A_{16}\left\{\frac{\partial u}{\partial x} + \frac{1}{2}\left(\frac{\partial w}{\partial x}\right)^2\right\}\right.\\
&\qquad\left. + A_{26}\left\{\frac{\partial v}{\partial y} + \frac{1}{2}\left(\frac{\partial w}{\partial y}\right)^2\right\} + A_{66}\left\{\frac{\partial u}{\partial y} + \frac{\partial v}{\partial x} + \frac{\partial w}{\partial x}\frac{\partial w}{\partial y}\right\}\right] = 0,\\
&\frac{\partial}{\partial y}\left[A_{12}\left\{\frac{\partial u}{\partial x} + \frac{1}{2}\left(\frac{\partial w}{\partial x}\right)^2\right\} + A_{22}\left\{\frac{\partial v}{\partial y} + \frac{1}{2}\left(\frac{\partial w}{\partial y}\right)^2\right\}\right.\\
&\qquad\left. + A_{26}\left\{\frac{\partial u}{\partial y} + \frac{\partial v}{\partial x} + \frac{\partial w}{\partial x}\frac{\partial w}{\partial y}\right\}\right] + \frac{\partial}{\partial x}\left[A_{16}\left\{\frac{\partial u}{\partial x} + \frac{1}{2}\left(\frac{\partial w}{\partial x}\right)^2\right\}\right.\\
&\qquad\left. + A_{26}\left\{\frac{\partial v}{\partial y} + \frac{1}{2}\left(\frac{\partial w}{\partial y}\right)^2\right\} + A_{66}\left\{\frac{\partial u}{\partial y} + \frac{\partial v}{\partial x} + \frac{\partial w}{\partial x}\frac{\partial w}{\partial y}\right\}\right] = 0.
\end{aligned}\right\} \tag{8.39}$$

Likewise, the equation of normal equilibrium is given by

$$\mathrm{L}_1 w = q + \frac{\partial^2 w}{\partial x^2}\left[A_{11}\left\{\frac{\partial u}{\partial x}+\frac{1}{2}\left(\frac{\partial w}{\partial x}\right)^2\right\}+A_{12}\left\{\frac{\partial v}{\partial y}+\frac{1}{2}\left(\frac{\partial w}{\partial y}\right)^2\right\}\right.$$
$$\left.+A_{16}\left\{\frac{\partial u}{\partial y}+\frac{\partial v}{\partial x}+\frac{\partial w}{\partial x}\frac{\partial w}{\partial y}\right\}\right]+2\frac{\partial^2 w}{\partial x \partial y}\left[A_{16}\left\{\frac{\partial u}{\partial x}+\frac{1}{2}\left(\frac{\partial w}{\partial x}\right)^2\right\}\right.$$
$$\left.+A_{26}\left\{\frac{\partial v}{\partial y}+\frac{1}{2}\left(\frac{\partial w}{\partial y}\right)^2\right\}+A_{66}\left\{\frac{\partial u}{\partial y}+\frac{\partial v}{\partial x}+\frac{\partial w}{\partial x}\frac{\partial w}{\partial y}\right\}\right]$$
$$+\frac{\partial^2 w}{\partial y^2}\left[A_{12}\left\{\frac{\partial u}{\partial x}+\frac{1}{2}\left(\frac{\partial w}{\partial x}\right)^2\right\}+A_{22}\left\{\frac{\partial v}{\partial y}+\frac{1}{2}\left(\frac{\partial w}{\partial y}\right)^2\right\}\right.$$
$$\left.+A_{26}\left\{\frac{\partial u}{\partial y}+\frac{\partial v}{\partial x}+\frac{\partial w}{\partial x}\frac{\partial w}{\partial y}\right\}\right], \tag{8.40}$$

where

$$\mathrm{L}_1 w = D_{11}\frac{\partial^4 w}{\partial x^4}+4D_{16}\frac{\partial^4 w}{\partial x^3 \partial y}+2(D_{12}+2D_{66})\frac{\partial^4 w}{\partial x^2 \partial y^2}$$
$$+4D_{26}\frac{\partial^4 w}{\partial x \partial y^3}+D_{22}\frac{\partial^4 w}{\partial y^4}.$$

Normally loaded plates

Substitution of (8.1) and (8.2) in (8.40) and equating terms of order Δ results in the small-deflexion equation (1.102) whose solution can be expressed in the form (8.5). Next, substitution of (8.1) and (8.2) into (8.39) and equating terms of order Δ^2 yields the following linear equations for determining u_2, v_2:

$$A_{11}\left\{\frac{\partial^2 u_2}{\partial x^2}+\frac{\partial w_1}{\partial x}\frac{\partial^2 w_1}{\partial x^2}\right\}+A_{26}\left\{\frac{\partial^2 v_2}{\partial y^2}+\frac{\partial w_1}{\partial y}\frac{\partial^2 w_1}{\partial y^2}\right\}$$
$$+(A_{12}+A_{66})\left\{\frac{\partial^2 v_2}{\partial x \partial y}+\frac{\partial w_1}{\partial y}\frac{\partial^2 w_1}{\partial x \partial y}\right\}+2A_{16}\left\{\frac{\partial^2 u_2}{\partial x \partial y}+\frac{\partial w_1}{\partial x}\frac{\partial^2 w_1}{\partial x \partial y}\right\}$$
$$+A_{16}\left\{\frac{\partial^2 v_2}{\partial x^2}+\frac{\partial w_1}{\partial y}\frac{\partial^2 w_1}{\partial x^2}\right\}+A_{66}\left\{\frac{\partial^2 u_2}{\partial y^2}+\frac{\partial w_1}{\partial x}\frac{\partial^2 w_1}{\partial y^2}\right\}=0, \tag{8.41}$$

$$A_{16}\left\{\frac{\partial^2 u_2}{\partial x^2}+\frac{\partial w_1}{\partial x}\frac{\partial^2 w_1}{\partial x^2}\right\}+A_{22}\left\{\frac{\partial^2 v_2}{\partial y^2}+\frac{\partial w_1}{\partial y}\frac{\partial^2 w_1}{\partial y^2}\right\}$$
$$+2A_{26}\left\{\frac{\partial^2 v_2}{\partial x \partial y}+\frac{\partial w_1}{\partial y}\frac{\partial^2 w_1}{\partial x \partial y}\right\}+(A_{12}+A_{66})\left\{\frac{\partial^2 u_2}{\partial x \partial y}+\frac{\partial w_1}{\partial x}\frac{\partial^2 w_1}{\partial x \partial y}\right\}$$
$$+A_{66}\left\{\frac{\partial^2 v_2}{\partial x^2}+\frac{\partial w_1}{\partial y}\frac{\partial^2 w_1}{\partial x^2}\right\}+A_{26}\left\{\frac{\partial^2 u_2}{\partial y^2}+\frac{\partial w_1}{\partial x}\frac{\partial^2 w_1}{\partial y^2}\right\}=0. \tag{8.42}$$

Similarly, substitution of (8.1) and (8.2) into (8.40) and equating terms of order Δ^3 now yields the following linear equation for determining w_3:

$$\begin{aligned}
L_1 w_3 = \alpha_3 &+ \frac{\partial^2 w_1}{\partial x^2}\left[A_{11}\left\{\frac{\partial u_2}{\partial x}+\frac{1}{2}\left(\frac{\partial w_1}{\partial x}\right)^2\right\}\right.\\
&\left.+A_{12}\left\{\frac{\partial v_2}{\partial x}+\frac{1}{2}\left(\frac{\partial w_1}{\partial y}\right)^2\right\}+A_{16}\left\{\frac{\partial u_2}{\partial y}+\frac{\partial v_2}{\partial x}+\frac{\partial w_1}{\partial x}\frac{\partial w_1}{\partial y}\right\}\right]\\
&+2\frac{\partial^2 w_1}{\partial x \partial y}\left[A_{16}\left\{\frac{\partial u_2}{\partial x}+\frac{1}{2}\left(\frac{\partial w_1}{\partial x}\right)^2\right\}+A_{26}\left\{\frac{\partial v_2}{\partial x}+\frac{1}{2}\left(\frac{\partial w_1}{\partial y}\right)^2\right\}\right.\\
&\left.+A_{66}\left\{\frac{\partial u_2}{\partial y}+\frac{\partial v_2}{\partial x}+\frac{\partial w_1}{\partial x}\frac{\partial w_1}{\partial y}\right\}\right]\\
&+\frac{\partial^2 w_1}{\partial y^2}\left[A_{12}\left\{\frac{\partial u_2}{\partial x}+\frac{1}{2}\left(\frac{\partial w_1}{\partial x}\right)^2\right\}+A_{22}\left\{\frac{\partial v_2}{\partial x}+\frac{1}{2}\left(\frac{\partial w_1}{\partial y}\right)^2\right\}\right.\\
&\left.+A_{26}\left\{\frac{\partial u_2}{\partial y}+\frac{\partial v_2}{\partial x}+\frac{\partial w_1}{\partial x}\frac{\partial w_1}{\partial y}\right\}\right].
\end{aligned} \tag{8.43}$$

The cycle of operations may, of course, be repeated.

8.4.2 *Strain energy in multi-layered plate*

Energy methods may also be used in the large-deflexion analysis of multi-layered anisotropic plates. In this connection we note that the analyses of Sections 1.8.1 and 6.2 maintain their validity in the large-deflexion régime because they are based simply on the relations between $\mathbf{M}, \mathbf{N}, \boldsymbol{\varepsilon}^0$ and $\boldsymbol{\kappa}$. Large-deflexion effects are introduced when we express $\boldsymbol{\varepsilon}^0$ in terms of the displacements u, v, w. Of the three expressions for the total strain energy per unit area U' given in Section 6.2, equation (6.16) is the most appropriate because it is expressed solely in terms of the plate properties and the strains and curvatures. Thus, collecting the relevant equations we have

$$U' = \tfrac{1}{2}(\boldsymbol{\varepsilon}^{0,\mathrm{T}}\mathbf{A}\boldsymbol{\varepsilon}^0 + 2\boldsymbol{\kappa}^{\mathrm{T}}\mathbf{B}\boldsymbol{\varepsilon}^0 + \boldsymbol{\kappa}^{\mathrm{T}}\mathbf{D}\boldsymbol{\kappa}). \tag{8.44}$$

where

$$\boldsymbol{\varepsilon}^0 = \left[\frac{\partial u}{\partial x}+\frac{1}{2}\left(\frac{\partial w}{\partial x}\right)^2, \frac{\partial v}{\partial y}+\frac{1}{2}\left(\frac{\partial w}{\partial y}\right)^2, \frac{\partial u}{\partial y}+\frac{\partial v}{\partial x}+\frac{\partial w}{\partial x}\frac{\partial w}{\partial y}\right]^{\mathrm{T}},$$

and

$$\boldsymbol{\kappa} = -\left[\frac{\partial^2 w}{\partial x^2}, \frac{\partial^2 w}{\partial y^2}, 2\frac{\partial^2 w}{\partial x \partial y}\right]^{\mathrm{T}}.$$

Note that the third term in (8.44) is given by (6.20) and the second term vanishes when there is zero coupling between **N** and **M**.

References

Chien, W. Z. Large deflection of a circular clamped plate under uniform pressure. *Chinese Phys.*, **7**, pp. 102–13 (1947).

Cox, H. L. Buckling of thin plates in compression. *Aero. Res. Council R. & M.* No. 1554. H.M.S.O. (August 1933).

Floor, W. K. G., and Burgerhout, T. J. Evaluation of the theory on the post-buckling behaviour of stiffened, flat, rectangular plates subjected to shear and normal loads. *N.L.L.* (Amsterdam), *Rep.* No. S. 370 (1951).

Koiter, W. T. The effective width of infinitely long flat rectangular plates under various conditions of edge restraint (in Dutch). *N.L.L.* (Amsterdam) *Rep.* No. S. 287 (December 1943).

——. Theoretical investigation of the diagonal tension field of flat plates (in Dutch). *N.L.L.* (Amsterdam) *Rep.* No. S. 295 (1944).

Nash, W. A., and Cooley, I. D. Large deflections of a clamped elliptical plate subjected to uniform pressure. *J. Appl. Mech.*, Ser. E, **26**, No. 2.

Stein, M. Loads and deformations of buckled rectangular plates. *N.A.S.A. Tech. Rep.* No. R-40 (March 1959).

Southwell, R. V., and Skan S. W. Stability under shearing forces of a flat elastic strip. *Proc. Roy. Soc.* Ser. A, **105**, p. 582 (1924).

Van der Neut, A., and Floor, W. K. G. Buckling behaviour of flat plates loaded in shear and compression. *N.L.L.* (Amsterdam) *Rep.* No. S 341 (1948).

Additional references

Hemp, W. S. The theory of flat panels buckled in compression. *Aero. Res. Council R. and M.* No. 2178. H.M.S.O. (1945).

Leggett, D. M. A. The stresses in a flat panel under shear when the buckling load has been exceeded. *Aero. Res. Council R. and M.* No. 2430. H.M.S.O. (1950).

Mansfield, E. H. On the post-buckling behaviour of stiffened plane sheet under shear. *Aero. Res. Council R. and M.* No. 3073. H.M.S.O. (1958).

Wang, Chi-Teh. Nonlinear large-deflection boundary-value problems of rectangular plates. *N.A.C.A. Tech. Note.* No. 1425 (March 1948).

——. Bending of rectangular plates with large deflections. *N.A.C.A. Tech. Note* No. 1462 (April 1948).

9

Asymptotic large-deflexion theories for very thin plates

The exact large-deflexion analysis of plates generally presents considerable difficulties, but there are three classes of plate problems for which simplified theories are available for describing their behaviour under relatively high loading. These 'asymptotic' theories are *membrane* theory, *tension field* theory (sometimes called *wrinkled membrane* theory) and *inextensional* theory. All are described below. For a plate of perfectly elastic material, the error involved in using these theories tends to zero as the loading is increased or as the thickness is reduced. In any practical material, however, there is a limit to the elastic strain that may be developed, and this in turn limits the range of validity of these asymptotic theories to plates which are very thin. For steel and aluminium alloys, a typical limit to the elastic strain is 0.004, and this restricts the range of validity of the asymptotic theories as follows. For *membrane theory* and *tension field* theory the thickness must be less than about 0.001 of a typical planar dimension, while for *inextensional* theory the thickness must be less than about 0.01 of a typical planar dimension.

9.1 **Membrane theory** (considered by Föppl 1907)

When a thin plate is continuously supported along the boundaries in such a manner that restraint is afforded against movement in the plane of the plate, the load tends to be resisted to an increasing extent by middle-surface forces. If the plate is sufficiently thin and the loads sufficiently high, the plate acts as a membrane whose flexural rigidity may be assumed to be zero. For the practically important case of the plate (membrane) of constant thickness without any temperature effects, the governing differential equations (7.1) and (7.3) then become

$$\left.\begin{aligned} q+\Diamond^4(\Phi,w)&=0\\ \nabla^4\Phi+\tfrac{1}{2}Et\Diamond^4(w,w)&=0.\end{aligned}\right\}\qquad(9.1)$$

It must be admitted that the solution of these equations still presents formidable difficulties, but certain problems which are mathematically one-dimensional admit of simple solutions, and these are now considered.

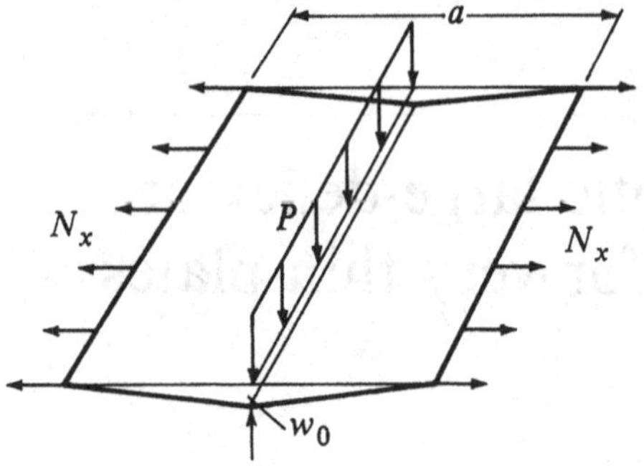

Fig. 9.1

9.1.1 Cylindrical deflexion of long membrane strip

Consider a long strip, of width a and constant thickness t, subjected to a load distribution that does not vary along its length. The edge supports are assumed to be either rigid or elastically restrained against movement in the plane of the membrane, as in Section 7.2. If, for example, the strip supports a central line load P per unit length (Fig. 9.1) it may readily be shown that, using membrane theory, the central deflexion w_0 and the middle-surface force per unit length N_x are given by

$$\left.\begin{aligned} w_0 &= \frac{a}{2}\left[P\left(\frac{\beta}{Et}+\frac{1}{aK}\right)\right]^{1/3} \\ \text{and}\qquad & \\ N_x &= \tfrac{1}{2}P^{2/3}\left(\frac{\beta}{Et}+\frac{1}{aK}\right)^{1/3}, \end{aligned}\right\} \tag{9.2}$$

where β and K are as defined in Section 7.2.1.

Similarly, for a uniform load q_0 it may be shown that

$$\left.\begin{aligned} w_0 &= \frac{a}{4}\left[3aq_0\left(\frac{\beta}{Et}+\frac{1}{aK}\right)\right]^{1/3}, \\ N_x &= \tfrac{1}{6}(3aq_0)^{2/3}\left(\frac{\beta}{Et}+\frac{1}{aK}\right)^{-1/3}, \end{aligned}\right\} \tag{9.3}$$

results that could also have been deduced from the analysis of Section 7.2 by letting $\eta \to \infty$.

9.1.2 Annular membrane under axial load

Consider an initially unstressed annular membrane whose outer boundary at r_2 is fixed and whose inner boundary at r_1 is attached to a floating rigid boss to which an axial load P is applied. The deflexion of the membrane is accordingly independent of θ and (9.1) may be written as

$$\left.\begin{aligned} &\frac{\mathrm{d}}{\mathrm{d}r}\left(\frac{\mathrm{d}\Phi}{\mathrm{d}r}\frac{\mathrm{d}w}{\mathrm{d}r}\right)=0,\\ &\frac{\mathrm{d}}{\mathrm{d}r}\left[r\frac{\mathrm{d}}{\mathrm{d}r}\left\{\frac{1}{r}\frac{\mathrm{d}}{\mathrm{d}r}\left(r\frac{\mathrm{d}\Phi}{\mathrm{d}r}\right)\right\}+\tfrac{1}{2}Et\left(\frac{\mathrm{d}w}{\mathrm{d}r}\right)^2\right]=0.\end{aligned}\right\} \tag{9.4a}$$

We search for a solution in the form

$$w(r)=k(r_2^\alpha-r^\alpha), \tag{9.4b}$$

which satisfies the condition that $w(r_2)=0$, and

$$\Phi(r)=cr^\beta, \tag{9.4c}$$

where k, c, α, β are constants.

Substitution into (9.4a) yields

$$\alpha\beta ckr^{\alpha+\beta-2}=\text{constant}, \tag{9.4d}$$

and

$$c\beta^2(\beta-2)r^{\beta-2}+\tfrac{1}{2}Etk^2\alpha^2r^{2\alpha-2}=\text{constant}. \tag{9.4e}$$

It follows from a consideration of the indices in (9.4d) and (9.4e) that

$$\alpha=2/3, \quad \beta=4/3. \tag{9.4f}$$

These values of α, β mean that (9.4d) is automatically satisfied while (9.4e) also requires that

$$c=\tfrac{3}{16}Etk^2. \tag{9.4g}$$

Now, the stress resultants in the radial and circumferential directions are

$$N_r=\frac{1}{r}\frac{\mathrm{d}\Phi}{\mathrm{d}r}, \quad N_\theta=\frac{\mathrm{d}^2\Phi}{\mathrm{d}r^2}, \tag{9.4h}$$

and hence equilibrium of the load P yields

$$\begin{aligned} P&=-2\pi rN_r\frac{\mathrm{d}w}{\mathrm{d}r}\\ &=\frac{16\pi}{9}ck, \end{aligned} \tag{9.4i}$$

in virtue of equations (9.4b), (9.4c) and (9.4f).

It follows from (9.4g) and (9.4i) that

$$\left.\begin{aligned} k&=\left(\frac{3P}{\pi Et}\right)^{1/3},\\ c&=\frac{3}{16}\left(\frac{9EtP^2}{\pi^2}\right)^{1/3}.\end{aligned}\right\} \tag{9.4j}$$

It is also necessary to consider the radial displacements u at the inner and outer boundaries. In this connection we note that

$$u = r\varepsilon_\theta, \tag{9.4k}$$

and the circumferential strain is given by

$$\varepsilon_\theta = \frac{1}{Et}(N_\theta - \nu N_r), \tag{9.4l}$$

where ν is the Poisson ratio.

It follows from equations (9.4c), (9.4f), (9.4h), (9.4k) and (9.4l) that

$$u = \frac{4cr^{1/3}}{3Et}\left(\tfrac{1}{3} - \nu\right), \tag{9.4m}$$

and the vanishing of u at the boundaries is seen to be fortuitously satisfied if $\nu = \frac{1}{3}$, which is typical of many elastic materials.

The solution to this problem was first given by Schwerin (1929), while the corresponding case with pre-tension was considered by Jahsman, Field and Holmes (1962).

9.1.3 Power law variation with load

The examples given in Sections 9.1.1 and 9.1.2 show that the deflexion at any point varies as (the loading)$^{1/3}$, and the middle-surface forces vary as (the loading)$^{2/3}$. These power law variations are a general property of initially unstressed membranes, as may be proved as follows.

Suppose that a particular solution of (9.1) is given by

$$q = q(x, y), \quad \Phi = \Phi(x, y), \quad w = w(x, y),$$

then it may be verified by substitution that another solution of (9.1) is given by

$$q' = \lambda q(x, y), \quad \Phi' = \lambda^{2/3}\Phi(x, y), \quad w' = \lambda^{1/3} w(x, y),$$

where λ is here a non-dimensional load parameter. This argument is restricted to initially unstressed membranes for which $\Phi = 0$ when $q = 0$.

9.1.4 Membranes with pre-tension

When the boundary supports impart middle-surface tensions throughout a membrane, as in a drum, a normal loading is resisted primarily by these tensions and the contribution of the flexural rigidity may be ignored. Thus under *small* deflexions the governing equations are adequately given by

$$\left.\begin{aligned} \nabla^4\Phi &= 0, \\ q + \lozenge^4(\Phi, w) &= 0. \end{aligned}\right\} \tag{9.5}$$

In particular, if the middle-surface forces are constant throughout the membrane we have:

$$q + N_x \frac{\partial^2 w}{\partial x^2} + 2N_{xy} \frac{\partial^2 w}{\partial x\, \partial y} + N_y \frac{\partial^2 w}{\partial y^2} = 0. \tag{9.6}$$

As a simple example, consider the fundamental mode of vibration of a rectangular membrane with sides a, b subjected to tensile forces N_x, N_y. The governing equation is (9.6) with q replaced by $-m\partial^2 w/\partial t^2$, where m is the mass per unit area of the membrane. The fundamental mode is of the form

$$w = w_0 \sin\frac{\pi x}{a} \sin\frac{\pi y}{b} \sin \Omega t,$$

where Ω is the circular frequency. Hence we find

$$\Omega^2 = \frac{\pi^2}{m}\left(\frac{N_x}{a^2} + \frac{N_y}{b^2}\right).$$

9.1.5 *Existence of other asymptotic states*

Membrane theory may also be used to infer the existence of other asymptotic states for membranes that are loaded only in their plane. Thus if q is zero, the first equation of (9.1) becomes

$$\lozenge^4(\Phi, w) = 0, \tag{9.7}$$

and this equation has three distinct forms of solution. Two of these, which we now consider, are mathematically trivial.

First, it is obvious that one form of solution is given by

$$w = 0,$$

for which the second equation of (9.1) yields

$$\nabla^4 \Phi = 0.$$

These are simply the equations of plane stress. The fact that the membrane remains flat, however, implies that neither of the principal stresses throughout the membrane can be compressive.

Similarly, the second form of solution of (9.7) is given by

$$\Phi = 0.$$

This means that there are no middle-surface stresses in the membrane, and this form of solution describes the state of a membrane subjected to 'compressive strains'.

The third, non-trivial, form of solution of (9.7) expresses the fact that

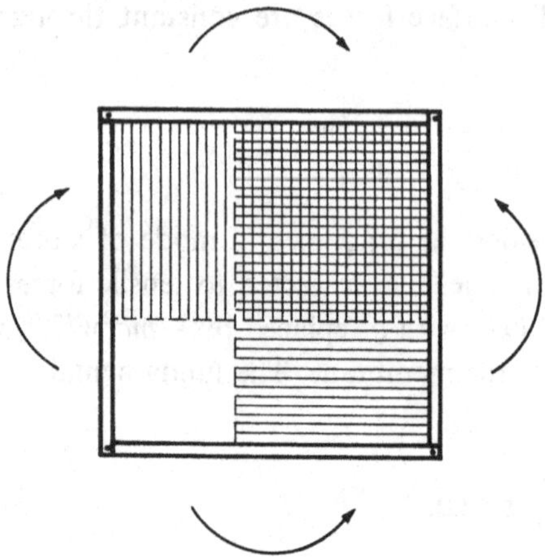

Fig. 9.2

at any point in the membrane one of the principal middle-surface stresses is zero, and the out-of-plane curvature in the direction of the other (tensile) principal stress is zero. This means that the only stresses in the membrane are tensile stresses carried along straight lines, or *tension rays*, stretching across the membrane from boundary to boundary. The membrane is envisaged as being finely wrinkled along normals to these tension rays. These features describe what is known as a *tension field*. The equations of membrane theory, however, do not lend themselves to the detailed solution of such problems. For this we require *tension field theory* (sometimes called *wrinkled membrane theory*) which is considered in Section 9.2. First, however, we note that it is possible for two, or even three, of the asymptotic membrane states to occur simultaneously. A simple case exhibiting all three states is the square plate supported by edge members to which moments are applied, as shown in Fig. 9.2, where the different membrane states are indicated by self-explanatory shading.

9.2 Tension field theory

Tension field theory describes the highly buckled (wrinkled) state of membranes or very thin plates whose boundaries are subjected to certain planar displacements well in excess of those necessary to initiate buckling. The theory was conceived by Wagner (1929) whose primary concern was to explain the behaviour of thin metal webs in beams and spars carrying a shear load well in excess of the initial buckling value. Such webs offer little resistance to the compressive strain component of the shear, and the spar flanges must be held apart by struts to prevent collapse. In the simple case of rigid spar flanges and rigid perpendicular struts, the stress field in the web in the highly buckled state is primarily that of tension at 45°. As the shear load increases, so does the magnitude of this tensile stress field and, just as a taut string resists a kinking action, so too does this tensile stress field resist the out-of-plane displacements engendered by the buckling action of the compressive stresses; these opposing actions result in a decreasing wavelength along the compressive buckles which form at right angles to the tension field. Strictly speaking, such problems are non-linear and their exact analysis presents formidable difficulties. However, for large values of the ratio (applied shear strain)/(shear strain at initial buckling) the flexural stresses and the planar compressive (post-buckling) stresses are negligible compared with the tensile stresses. In tension field theory the simplifying assumption is made that these relatively negligible stresses are zero, which is physically equivalent to the assumption of zero flexural membrane stiffness: the membrane is envisaged as being finely wrinkled at right angles to the lines of tension. In general these *tension rays* are not necessarily parallel and the boundary conditions need not be those of pure shear, as in the previous example, but shear must play a dominant role in the boundary deformation because of the requirement that the principal strains at any point are of opposite sign.

In his original paper Wagner presented a method for determining the distribution of tension rays in the general case, but this was based on lengthy geometrical considerations. Reissner (1938) achieved a simpler analysis based on straightforward calculus, while the concept of a variable Poisson ratio, with particular reference to the partly wrinkled membrane, was introduced by Stein and Hedgepeth (1961). In the specific problems which these writers solve the tension rays exhibit a repetitive pattern.

These authors and the writer were unaware that important advances in tension field theory had been made in Japan. Kondo (1938) had independently derived a method of analysis similar to that of Reissner and had also presented the first exact solutions to problems involving a non-repetitive pattern of rays. Iai (1943) had shown how the distribution of the tension rays could be determined by focusing attention solely on the displacement component along the tension rays; the method is based on a principle of maximum strain energy under given boundary displacements. This, and subsequent Japanese work on curved tension fields, is well documented in a review paper (in English) by Kondo, Iai, Moriguti and Murasaki (1955).

The writer's association with tension field theory began accidentally and unknowingly with the development of the *inextensional theory* (1955) for describing the large-deflexion *bending* behaviour of certain plates. The association with tension field theory became apparent only in 1968 with the realization that inextensional theory and tension field theory are analogous and have dual properties similar to those exhibited by small-deflexion plate theory and plane stress theory. Indeed, detailed inextensional solutions by Mansfield and Kleeman (1955b) for tip loaded triangular cantilever plates, including contour lines of maximum principal stress, are directly applicable to dual tension field problems. In the context of earlier work on tension fields the writer's (dual) analysis links the principle of maximum strain energy to a coordinate system related to the membrane geometry via a coordinate system related to the (unknown) distribution of tension rays. It thus combines in a concise but distinct form features first introduced by Iai, Kondo and Reissner.

9.2.1 Analysis

We consider an initially flat membrane of arbitrary shape and variable stiffness. The membrane is loaded only at its boundaries which are subjected to given planar displacements or, if a boundary is straight, either the displacements are given or the edge is free. The boundary conditions generate a tension field over the entire membrane, an assumption which, for topological reasons, restricts the number of free edges that can be considered to two or less.

The lines of principal stress necessarily form an orthogonal curvilinear network and it is convenient to consider the equilibrium of an infinitesimal curvilinear rectangle bounded by such lines. Equilibrium in the direction of the zero principal stress gives

$$t\sigma_\eta \varkappa = 0, \tag{9.8}$$

where σ_η is the non-zero (tensile) principal stress and $\varkappa$ is the curvature of this principal stress trajectory. Accordingly, $\varkappa$ is zero and the trajectories of the tensile principal stresses are straight lines, hereafter referred to as *tension rays* (see Fig. 9.3).

Equilibrium in the direction of a tension ray demands continuity of the tensile load carried between adjacent rays, and hence along any ray

$$\eta t\sigma_\eta = \text{constant}, \tag{9.9}$$

where η is the distance from the point of intersection of adjacent rays. Of course, if the tension rays are parallel, (9.9) reduces to

$$t\sigma_\eta = \text{constant}. \tag{9.10}$$

9.2.2 The principle of maximum strain energy

The fundamental problem of tension field theory lies in determining the orientation of the tension rays. In this connection we note again that there are no direct and shear stresses across adjacent rays; accordingly, the strain energy of the membrane is solely due to tensile stresses directed along the

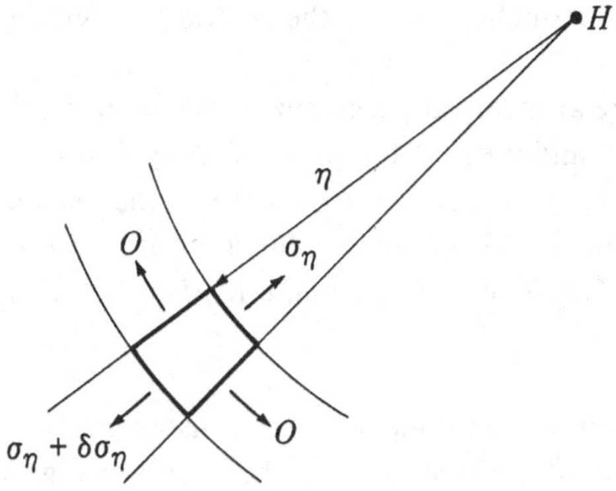

Fig. 9.3

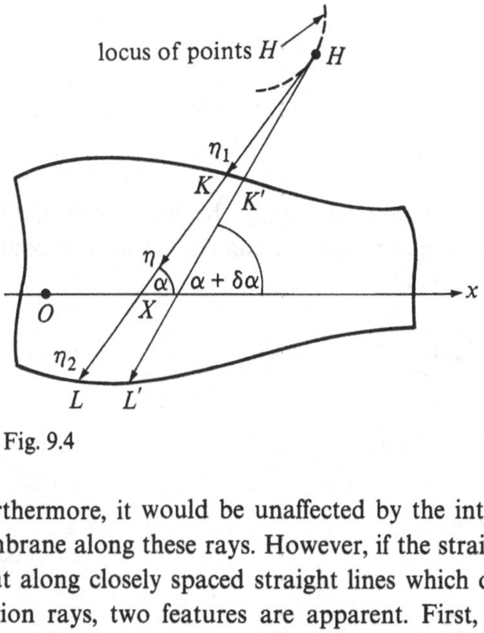

Fig. 9.4

rays; furthermore, it would be unaffected by the introduction of cuts in the membrane along these rays. However, if the strained membrane were to be cut along closely spaced straight lines which do *not* coincide with the tension rays, two features are apparent. First, the resultant strain energy would be solely due to tensile stresses directed along these lines and, second, this strain energy would be *less* than in the uncut membrane because, with no transfer of energy from the boundaries, successive cuts could only release energy. It follows that the true distribution of tension rays maximizes this 'tensile' strain energy, and it is this principle which forms the basis of the following variational analysis. Fig. 9.4 shows a membrane with a reference axis Ox; a typical tension ray cuts the x-axis at a point X and at an angle α, so that we can write in a formal manner

$$\alpha = \alpha(X). \tag{9.11}$$

The determination of the function $\alpha(X)$ is the central problem in tension field theory.

The lines LK, $L'K'$ are arbitrary adjacent cuts which intersect the x-axis at angles α and $(\alpha + \delta\alpha)$ and meet at the point H. A typical point in the membrane is specified by coordinates α, η, where η is the distance along the α-line from the point H which is itself a function of α and X.

The elemental slice $KLL'K'$ undergoes *radial tension* such that

$$\eta E t \varepsilon_\eta = c_\alpha, \text{ a constant,} \tag{9.12}$$

where ε_η is the direct strain along the line LK. The constant c_α is determined by equating the line-integral of the strain ε_η to the known change in length of LK, denoted by Δ_α:

$$\int_{\eta_1}^{\eta_2} \varepsilon_\eta \, \mathrm{d}\eta = \Delta_\alpha, \tag{9.13}$$

whence,

$$c_\alpha = \frac{\Delta_\alpha}{\displaystyle\int_{\eta_1}^{\eta_2} \frac{1}{\eta E t} \mathrm{d}\eta}. \tag{9.14}$$

The strain energy of the membrane will now be determined, and in doing so it is convenient to integrate first over an elemental slice bounded by adjacent cuts. Thus

$$\begin{aligned} U &= \frac{1}{2} \iint E t \varepsilon_\eta^2 \, \mathrm{d}A \\ &= \frac{1}{2} \int \int_{\eta_1}^{\eta_2} E t \varepsilon_\eta^2 \eta \, \mathrm{d}\eta \, \mathrm{d}\alpha \\ &= \frac{1}{2} \int F \, \mathrm{d}X \end{aligned} \tag{9.15}$$

where

$$F = \frac{\alpha' \Delta_\alpha^2}{\displaystyle\int_{\eta_1}^{\eta_2} \frac{1}{\eta E t} \mathrm{d}\eta}, \tag{9.16}$$

and a prime denotes differentiation with respect to X. Now from geometrical considerations the value of η at the x-axis is given by

$$\eta_X = \pm \frac{\sin \alpha}{\alpha'}, \tag{9.17}$$

(the sign depending on whether H is above or below the x-axis) and it follows that F is a known function of X, α, α' and the boundary conditions.

The function $\alpha(X)$ is to be determined from the condition that the strain energy U is a maximum. It follows from the calculus of variations that

$$\alpha'' F_{\alpha'\alpha'} + \alpha' F_{\alpha\alpha'} + F_{X\alpha'} - F_{\alpha} = 0 \tag{9.18}$$

in which F_{α}, for example, stands for $\partial F/\partial\alpha$ where F is formally regarded as a function of independent variables X, α, α'. Equation (9.18) is a non-linear differential equation of the second order, and, with the boundary conditions, it determines the function $\alpha(X)$. The outline solution of (9.18) for a number of examples is given later, but first we draw attention to a simple but important class of problems, namely those in which F does not contain X explicitly; the equation may then be integrated once to give

$$F - \alpha' F_{\alpha'} = \text{constant}. \tag{9.19}$$

Also we note that for the practically important case in which Et is constant

$$\frac{F}{Et} = \frac{\alpha' \Delta_{\alpha}^2}{\ln(\eta_2/\eta_1)}. \tag{9.20}$$

Uniform strain field

If the displacements along the complete boundary are consistent with a uniform strain field, the tension rays are parallel and coincide with the direction of the positive principal strain; this direction may be determined by maximizing $\Delta_{\alpha}/(\eta_2 - \eta_1)$.

9.2.3 *Validity of solutions*

Ideally, any theoretical solution should be checked to see that the requirements for a tension field have not been violated. Thus it should be shown that throughout the membrane

$$\Delta_{\alpha} > 0 \tag{9.21}$$

and

$$\varepsilon_{\alpha} + \nu\varepsilon_{\eta} \leqslant 0, \tag{9.22}$$

where ε_{α} is the strain along a normal to a tension ray. A check on (9.21) is relatively straightforward but a thorough check on (9.22) is not only time-consuming but also frustrating because if the inequality is not satisfied – an indication that the membrane is not completely wrinkled – it is, nevertheless, virtually impossible to obtain a more meaningful solution. Indeed, the approximate nature of tension field theory itself does not justify too scrupulous an attention to detail. For these reasons we suggest that a check on the inequality (9.22) be confined to the boundaries where violations, if any, are likely to occur, and where ε_{α} assumes the following simple form:

$$\varepsilon_\alpha = \varepsilon_s \operatorname{cosec}^2 \lambda - \varepsilon_\eta \cot^2 \lambda,$$

where ε_s is the strain along the boundary and λ is the angle that the tension ray makes with a tangent to the boundary. Thus at the boundary the inequality assumes the form

$$\varepsilon_s \operatorname{cosec}^2 \lambda + \varepsilon_\eta(\nu - \cot^2 \lambda) \leqslant 0. \tag{9.23}$$

For example, if ε_s is zero the inequality requires that

$$\cot^2 \lambda \geqslant \nu,$$

that is,

$$\left.\begin{aligned} &\lambda \leqslant 63\tfrac{1}{2}^\circ \\ \text{or}\quad & \\ &\lambda \geqslant 116\tfrac{1}{2}^\circ \end{aligned}\right\} \quad \text{if } \nu = \tfrac{1}{4}.$$

9.2.4 *Shearing of semi-infinite membrane strip*

Consider the semi-infinite parallel strip of width a, whose opposite edges undergo a constant shear displacement U_0, as shown in Fig. 9.5. We will obtain the solution for the case of a free edge at $\alpha_0 = \frac{1}{2}\pi$, but it should be realized that this necessarily contains the solution for $\alpha_\infty < \alpha_0 < \frac{1}{2}\pi$; indeed, it also contains the solution for $\alpha_0 > \frac{1}{2}\pi$, for the additional triangular region beyond $\alpha_0 = \frac{1}{2}\pi$ is unstressed. (Strictly speaking, there are limitations on the value of α_0, as discussed above, but from a practical standpoint this restriction is not of much significance.)

With the notation of Fig. 9.5, it is seen that α' is negative so that the locus of points H lies below the x-axis and accordingly $\eta_1 > \eta_2$. Thus

$$\left.\begin{aligned} \Delta_\alpha &= U_0 \cos \alpha, \\ \eta_1 - \eta_2 &= a \operatorname{cosec} \alpha, \\ \eta_X = \eta_2 &= -\frac{\sin \alpha}{\alpha'}, \end{aligned}\right\} \tag{9.24}$$

so that

$$F \propto \frac{\alpha' \cos^2 \alpha}{\ln(1 - a\alpha' \operatorname{cosec}^2 \alpha)}. \tag{9.25}$$

[Note that a pre-knowledge of the sign of α' is not essential to the analysis. Thus if α' had been assumed positive, the only alteration in the expression for F is an overall sign change, and this itself does not affect the form of (9.25).]

At this stage it is convenient to introduce

$$\mu = \eta_1/\eta_2 = 1 - a\alpha' \operatorname{cosec}^2 \alpha, \quad (\text{i.e., } \mu \geqslant 1) \tag{9.26}$$

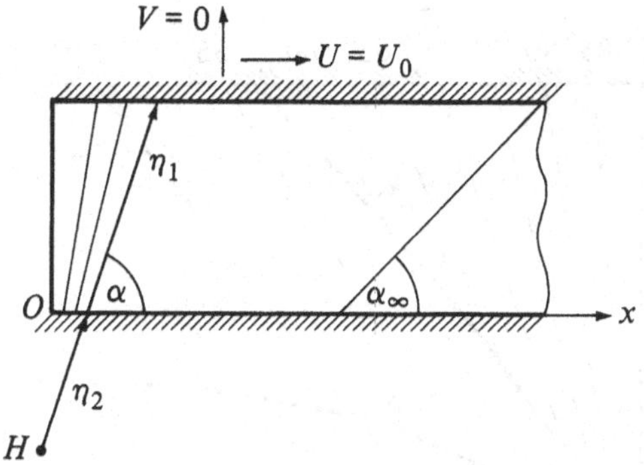

Fig. 9.5

in terms of which (9.25) and (9.19) become

$$\sin^2 2\alpha = C\mu\left(\frac{\ln\mu}{\mu-1}\right)^2. \tag{9.27}$$

The constant of integration is determined from the condition that as $x\to\infty$ the tension rays become parallel and $\mu\to 1$, while $\alpha\to\frac{1}{4}\pi$, the value which maximizes $\Delta_\alpha/(\eta_1-\eta_2)$. Hence $C=1$ and the relation between α' and α is readily determined numerically from (9.26) and (9.27); the α, X relation is obtained by integration.

The stresses are given by (9.12) and (9.14). As $x\to\infty$ it may be shown that

$$\sigma_{\eta,\infty} = EU_0/2a \tag{9.28}$$

and it is convenient to introduce

$$\sigma = \sigma_\eta/\sigma_{\eta,\infty}, \tag{9.29}$$

which is now a non-dimensional measure of the stress and the equivalent of a stress concentration factor. The peak values of σ, denoted by σ^*, occur along the x-axis ($\eta=\eta_2$), where it may be shown that they satisfy the relation

$$\sin 2\alpha = \frac{2\sigma^*\ln\sigma^*}{\sigma^{*2}-1}. \tag{9.30}$$

Contours of constant σ and tension ray lines at 5° intervals are shown in Fig. 9.6 together with a shaded region which indicates the unstressed triangular zone which occurs when $\alpha_0>\frac{1}{2}\pi$. Solutions for strips of finite length are obtained by taking different values of the constant C. In practice, of course, the opposite edges of the strip tend to be pulled together by the action of the membrane stresses, and the condition that $V=0$ is only possible if these edges are held apart in a rigid manner; when they are

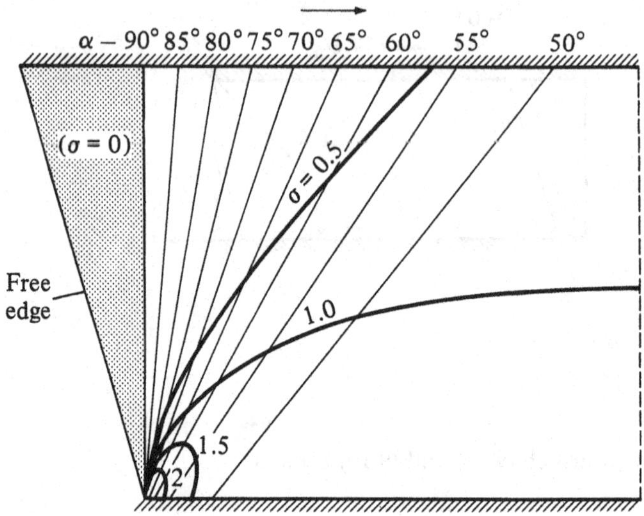

Fig. 9.6 Tension ray lines and stresses in semi-infinite membrane strip under shear ($\alpha_0 > \pi/4$).

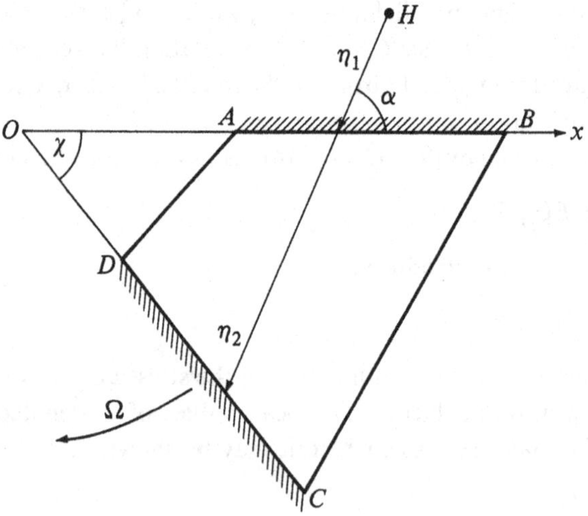

Fig. 9.7 Quadrilateral membrane.

held apart elastically some contraction occurs. Such problems have been discussed elsewhere (Mansfield 1968).

9.2.5 A quadrilateral membrane

Here we consider an arbitrary quadrilateral membrane *ABCD* with one pair (*AD*, *BC*) of opposite edges free and the other pair subjected to a relative rotation Ω about their point of intersection *O*, as shown in Fig. 9.7.

The problem demonstrates the integration of (9.18) rather than the simpler (9.19). At the same time, it exhibits the simplifying feature that any solution necessarily generates further solutions by a linear scaling of dimensions. Mathematically, this manifests itself in the fact that (9.19) can be expressed as a first-order differential equation in $X\alpha'$ which readily lends itself to numerical integration.

With the notation of Fig. 9.7,

$$\Delta_\alpha = \Omega X \sin\alpha, \tag{9.31}$$

while from geometrical considerations

$$\left.\begin{aligned} \eta_1 &= \frac{\sin\alpha}{\alpha'}, \\ \eta_2 &= \eta_1 + \frac{X\sin\chi}{\sin(\alpha+\chi)}. \end{aligned}\right\} \tag{9.32}$$

Thus, from (9.20),

$$\left.\begin{aligned} \frac{F}{Et\Omega^2} &= \frac{X^2\alpha'\sin^2\alpha}{\ln\mu}, \\ \text{where}\quad \mu &= \eta_2/\eta_1 \\ &= 1 + \frac{X\alpha'\sin\chi}{\sin\alpha\sin(\alpha+\chi)}. \end{aligned}\right\} \tag{9.33}$$

Substitution of (9.33) into (9.18) yields an equation of the second order in X and α which can, however, be expressed as one of the first order in $X\alpha'$ and α or, more conveniently, in μ and α:

$$\left.\begin{aligned} \frac{d\alpha}{d\mu} &= -\frac{\sin\alpha\sin(\alpha+\chi)}{p\sin(2\alpha+\chi)+q\sin\chi}, \\ \text{where}\quad p &= \frac{2\mu(\mu-1)\ln\mu}{(\mu+1)\ln\mu-2(\mu-1)}, \\ q &= \frac{\mu\ln\mu(\mu^2-1-2\mu\ln\mu)}{(\mu-1)\{(\mu+1)\ln\mu-2(\mu-1)\}}. \end{aligned}\right\} \tag{9.34}$$

This can be integrated numerically, assuming various starting values for $[\mu]_{\alpha=\alpha_0}$. Once this primary integration has been completed, the formal solution for X and α is straightforward. Thus, from the definition of μ in (9.33),

$$\frac{1}{X}\frac{dX}{d\alpha} = \frac{\sin\chi}{(\mu-1)\sin\alpha\sin(\alpha+\chi)},$$

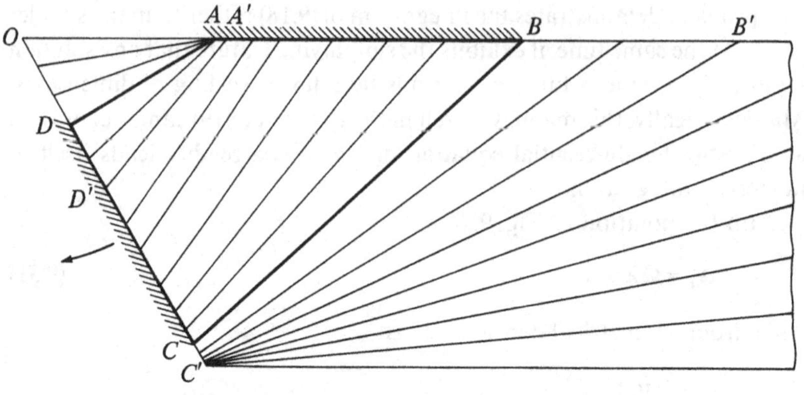

Fig. 9.8 Tension rays in quadrilateral membrane, $\mu_A = 50$.

which may be integrated to give

$$X = X_0 \exp\left\{ \int_{\alpha_0}^{\alpha} \frac{\sin \chi}{(\mu - 1)\sin\alpha \sin(\alpha + \chi)} \mathrm{d}\alpha \right\}. \tag{9.35}$$

The form of the solution depends markedly on whether or not μ passes through the value of unity, that is, on whether or not the point H passes from one side of the quadrilateral to the opposite side. In the former case care must be exercised as $\mu \to 1$ where it is preferable to replace the numerical integration by an analytical integration based on an expansion of the functions p, q about the point $\mu = 1$. The two forms of the solution are typified by Figs. 9.8, 9.9 which show the distribution of tension rays in quadrilateral membranes specified by

$$A\hat{O}D = \tfrac{1}{3}\pi (= \chi),$$
$$A\hat{D}C = \tfrac{1}{2}\pi, \quad \text{so that} \quad \alpha_0 = \tfrac{1}{6}\pi.$$

The position of the corner points B, C are such that

$$\left.\begin{aligned} AB/OA &= 1.60, \\ \mu_A &= 50 \end{aligned}\right\} \quad \text{in Fig. 9.8}$$

and

$$\left.\begin{aligned} AB/OA &= 0.68, \\ \mu_A &= 60 \end{aligned}\right\} \quad \text{in Fig. 9.9.}$$

The magnitudes of AB/OA were chosen arbitrarily. However, because any tension ray line can be regarded as a free edge, the solutions presented apply to an infinity of quadrilaterals with different ratios AB/OA and different angles $A\hat{D}C$ such as $A'B'C'D'$. It will be seen that all these solutions are embodied in the solution for two semi-infinite parallel strips with skew edges. The strips are such that the lines OAB are part free and

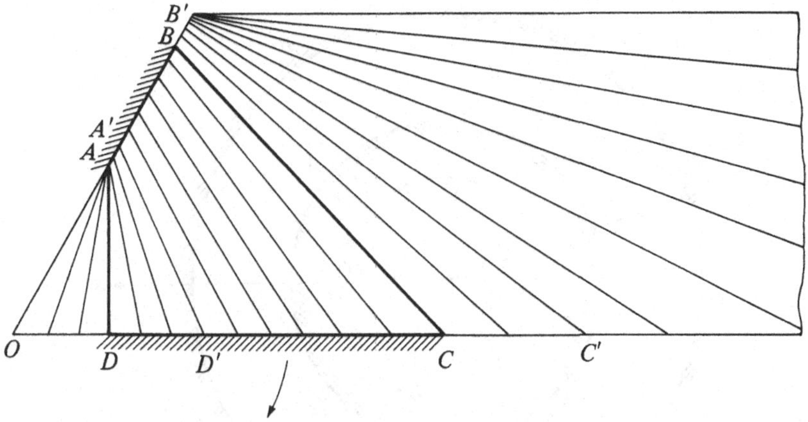

Fig. 9.9 Tension rays in quadrilateral membrane, $\mu_A = 60$.

part clamped while the lines *ODC* are entirely clamped; in Fig. 9.8, however, the line *OAB* is infinitely long and the line *ODC* is finite while the opposite occurs in Fig. 9.9.

Finally, we draw attention to the fact that in the quadrilaterals *ABCD* peaks of stress occur at points *A*, *C* in Fig. 9.8 and at points *A*, *B* in Fig. 9.9. At point *A*, for example,

$$\sigma_{\eta,A} = \frac{E\Omega(\mu_A - 1)\sin\alpha_A \sin(\alpha_A + \chi)}{\ln\mu_A \sin\chi}. \tag{9.36}$$

If this is expressed as a multiple of the stress in a vanishingly narrow *parallel* strip of membrane along the line *AD*, the resultant 'stress concentration factor' is given by

$$\left.\begin{aligned} S_A &= \frac{\mu_A - 1}{\ln\mu_A} \\ &= 12.5 \text{ for the quadrilateral in Fig. 9.8,} \\ &= 14.4 \text{ for the quadrilateral in Fig. 9.9.} \end{aligned}\right\} \tag{9.37}$$

These high values are to be expected in view of the convergent bunching of the tension rays near point *A*.

9.2.6 *Torsion of slit annular membrane*

Throughout the present analysis the angle α is measured from a fixed direction, namely the *x*-axis, and this feature enables us to use the area relation

$$\delta A = \eta\delta\alpha\,\delta\eta$$

in the derivation of (9.15). But the ordinate *X* is not an essential ingredient

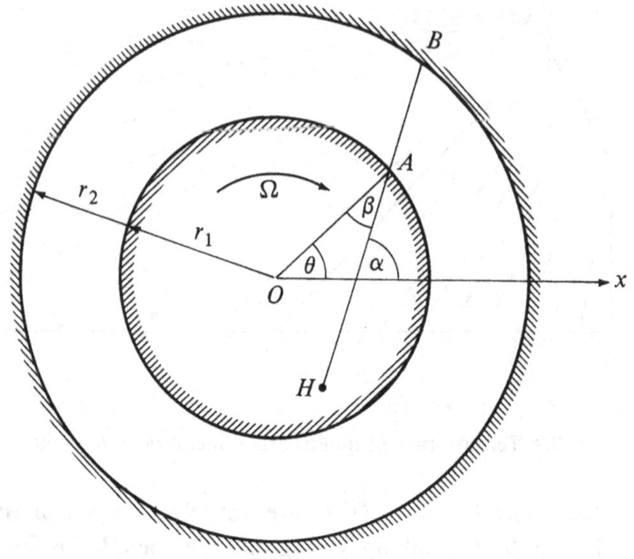

Fig. 9.10

in the analysis and some other reference system may be preferable. To demonstrate this we now consider a membrane bounded by concentric circles and a radial cut, the circular boundaries undergoing a relative rotation about their common centre. We adopt the notation of Fig. 9.10 in which a typical tension ray is specified by the angle β and by the angular position θ of its intersection with the inner circular boundary. It follows that

$$\alpha = \theta + \beta,$$

and (9.15), (9.20) may be written as

$$\left. \begin{aligned} U &= \frac{1}{2}\int F \, \mathrm{d}\theta, \\ \text{where} \quad & \\ F &= \frac{Et(1+\beta')\Delta_\alpha^2}{\ln(\eta_2/\eta_1)}, \end{aligned} \right\} \tag{9.38}$$

and a prime denotes differentiation with respect to θ. Now from geometrical considerations

$$\eta_\theta = HA = \frac{r_1 \cos\beta}{(1+\beta')}, \tag{9.39}$$

and

$$\left. \begin{aligned} AB &= r_1\{(k^2 - \sin^2\beta)^{1/2} - \cos\beta\} \\ \text{where} \quad & \\ k &= r_2/r_1. \end{aligned} \right\} \tag{9.40}$$

Thus

$$\left.\begin{aligned} \eta_2/\eta_1 &= \mu, \text{ say,} \\ &= 1 + (1+\beta')J(\beta), \\ \text{where} \quad & \\ J(\beta) &= \sec\beta(k^2 - \sin^2\beta)^{1/2} - 1. \end{aligned}\right\} \tag{9.41}$$

Also,

$$\Delta_\alpha = \Omega r_1 \sin\beta \tag{9.42}$$

and hence from (9.16)

$$\frac{F}{Et\Omega^2 r_1^2} = \frac{(1+\beta')\sin^2\beta}{\ln\{1 + (1+\beta')J(\beta)\}}. \tag{9.43}$$

This expression for F does not contain θ explicitly and (9.18) can therefore be integrated once to give

$$F - \beta' F_{\beta'} = CEt\Omega^2 r_1^2, \text{ say,} \tag{9.44}$$

whence

$$C\mu \ln^2\mu = \sin^2\beta\left\{\mu\ln\mu + (\mu - 1)\left(\frac{\mu-1}{J(\beta)} - 1\right)\right\}. \tag{9.45}$$

By letting C assume various values we obtain from (9.45) and (9.41) a relation between β and β' which may be integrated to yield the β, θ relation. Before considering a specific example, however, it is expedient to obtain the solution for the uncut annular membrane. In this case, $\beta' = 0$ and β is determined from the condition

$$F_\beta = 0,$$

whence

$$\ln\left(\frac{k^2 - \sin^2\beta}{\cos^2\beta}\right) = \frac{\tan^2\beta(k^2-1)}{k^2 - \sin^2\beta}, \tag{9.46}$$

which is in agreement with the solution obtained by Reissner (1938). The resulting 'steady-state' value of β, β^*, say, given by (9.46), may be used to obtain an adequate approximation to C in (9.45). This is because it is known that away from the immediate vicinity of a cut the angle β approaches the steady-state value β^*. Furthermore, if a cut along $\theta = 0$, say, is regarded as a cut at $\theta = 0$ and another at $\theta = 2\pi$, it follows from Saint Venant's principle that a deviation in the angle of the cut at $\theta = 0$, say, has virtually no effect on the pattern of rays near $\theta = 2\pi$, and vice versa. The value of the constant C appropriate to the pattern of rays on either side of the cut is thus adequately given by (9.45) with $\beta = \beta^*$ and β' zero, whence

$$C \approx \frac{1}{2}\left(\frac{k^2 - \sin^2\beta^*}{k^2 - 1}\right)^{1/2} \sin 2\beta^*. \tag{9.47}$$

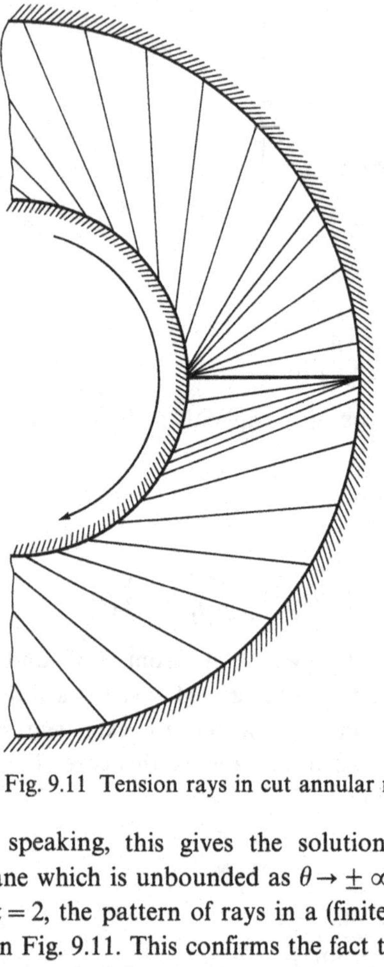

Fig. 9.11 Tension rays in cut annular membrane.

Strictly speaking, this gives the solution for a semi-infinite annular membrane which is unbounded as $\theta \rightarrow \pm \infty$.

For $k=2$, the pattern of rays in a (finite) annulus with a radial cut is shown in Fig. 9.11. This confirms the fact that away from the immediate neighbourhood of the cut the ray orientation angle β rapidly approaches the steady-state value appropriate to an uncut membrane. Furthermore, because any tension ray can be regarded as a cut, the solution effectively embodies the solution for an annular membrane with a single cut, or bounded by widely separated cuts, at any angle between zero and the steady-state value. Finally, attention is drawn to the fact that as θ approaches 2π, adjacent rays pass through a configuration in which they are parallel, leading to a constant value of the tensile stress along the ray; beyond this critical configuration the peak stresses occur at the *outer* boundary.

9.2.7 Parallel tension field in a spar web

As we have already noted, tension fields in the webs of I-beams were first considered by Wagner and in this context they are frequently referred to as *Wagner tension fields*. In many cases the generators are parallel – or

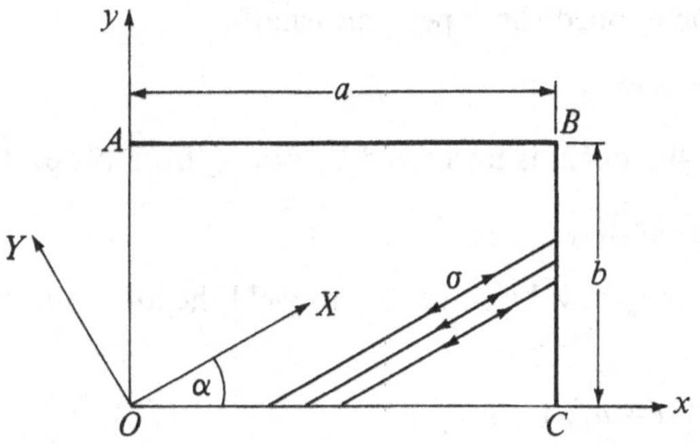

Fig. 9.12

may be assumed to be so – and the only problem is then to determine their orientation, which depends on the loading and the in-plane stiffness of the boundary supports.

Fig. 9.12 represents part of an I-beam under shear. The members AB, OC are the spars and AO, BC are vertical struts.

The angle α of the tension rays is now determined in terms of given constant strains ε_x and ε_y in the edge members and a constant shear strain ε_{xy} in the panel.

If the strains $\varepsilon_x, \varepsilon_y, \varepsilon_{xy}$ are related to new axes OX, OY parallel to and at right angles to the tension rays, we obtain

$$\left.\begin{aligned} \varepsilon_X &= \tfrac{1}{2}\varepsilon_x(1+\cos 2\alpha) + \tfrac{1}{2}\varepsilon_y(1-\cos 2\alpha) + \tfrac{1}{2}\varepsilon_{xy}\sin 2\alpha, \\ \varepsilon_Y &= \tfrac{1}{2}\varepsilon_x(1-\cos 2\alpha) + \tfrac{1}{2}\varepsilon_y(1+\cos 2\alpha) - \tfrac{1}{2}\varepsilon_{xy}\sin 2\alpha, \\ \varepsilon_{XY} &= (\varepsilon_y - \varepsilon_x)\sin 2\alpha + \varepsilon_{xy}\cos 2\alpha, \end{aligned}\right\} \quad (9.48)$$

and the angle α is to be determined from the condition that these are principal strains, so that

$$\tan 2\alpha = \frac{\varepsilon_{xy}}{\varepsilon_x - \varepsilon_y}. \quad (9.49)$$

The stress σ_X along the tension rays is now given by substituting this value of α into (9.48) to give

$$\begin{aligned} \sigma_X &= E\varepsilon_X \\ &= \tfrac{1}{2}E[\varepsilon_x + \varepsilon_y + \{\varepsilon_{xy}^2 + (\varepsilon_y - \varepsilon_x)^2\}^{1/2}]. \end{aligned} \quad (9.50)$$

Equilibrium conditions

If the section area of each of the members AB, OC in Fig. 9.12 is F_1 while that in AO, BC is F_2, we find on resolving in the x- and y-directions:

$$\left.\begin{aligned} 2F_1E\varepsilon_x + bt\sigma_X\cos^2\alpha = 0, \\ 2F_2E\varepsilon_y + at\sigma_X\sin^2\alpha = 0. \end{aligned}\right\} \quad (9.51)$$

Similarly, if S is the applied shear per unit length

$$S = t\sigma_X \sin\alpha\cos\alpha. \tag{9.52}$$

The solution of (9.49)–(9.52) is facilitated by noting from (9.50) that

$$\varepsilon_{xy} = 2\{(\varepsilon_X - \varepsilon_x)(\varepsilon_X - \varepsilon_y)\}^{1/2}. \tag{9.53}$$

These equations may now be combined to yield the following equation for determining α

$$\left.\begin{aligned} \tan 2\alpha &= \frac{2\{(1+n_1)(1+n_2)\}^{1/2}}{n_2 - n_1} \\ \text{where}\quad n_1 &= \frac{bt\cos^2\alpha}{2F_1}, \\ n_2 &= \frac{at\sin^2\alpha}{2F_2}. \end{aligned}\right\} \tag{9.54}$$

9.3 Inextensional theory (considered by Mansfield 1955)

A thin plate which is free to deflect along its entire periphery, or which is clamped along one straight boundary and free elsewhere, tends to resist an applied normal loading by its flexural rigidity alone. This is because the boundary conditions preclude the possibility of the formation of significant middle-surface forces. The simplifying assumption may now be made that the middle-surface strains are zero, and because of this the theory is referred to as *inextensional theory*. The basic assumption of inextensional theory is thus physically equivalent to the introduction of constraints into the plate, and it follows that all stiffnesses calculated on inextensional theory will be overestimates.

Now it is a well-known geometrical fact that an initially flat inextensible surface can deform only into a *developable* surface such that through every point it is possible to draw one straight line which lies entirely in the deflected surface. These lines are the *generators* of the deflected surface and the fundamental problem of inextensional theory lies in their determination. In some instances the positioning of the generators can be deduced immediately. For example, the generators lie along the radii in a shallow cone formed from an originally flat plate bounded by radii and concentric circles (Fig. 9.13); furthermore, all the generators here meet at a point, and the radius of curvature along a line normal to a generator is directly proportional to the distance from this point.

We shall later make use of this property of *conical bending* in discussing general inextensible deformation of plates. This is because each elemental slice of plate bounded by adjacent (non-parallel) generators undergoes conical bending, although the 'radius' and 'curvature' of each elemental

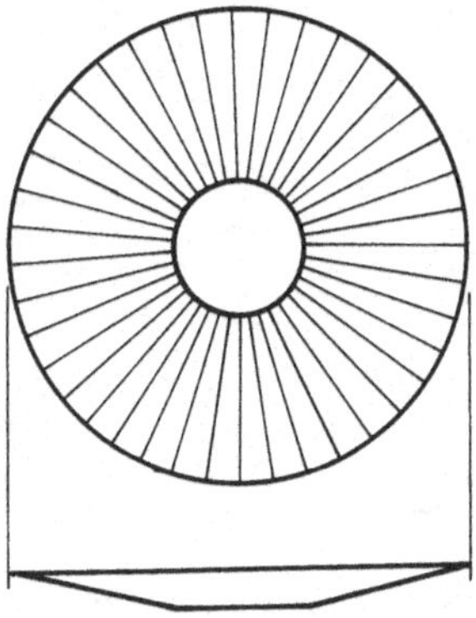

Fig. 9.13

conical slice vary continuously over the plate. The locus of points of intersection of adjacent generators is termed the *edge of regression.*

9.3.1 The principle of maximum strain energy

The fundamental problem of inextensional theory lies in determining the orientation of the generators. In this connection, we note that at any point in the plate the principle curvatures lie along and at right angles to the generators. The curvature along the generator is necessarily zero, and it follows that the strain energy is solely due to bending about the generators. Furthermore, the strain energy would be unaffected by the introduction of (hypothetical) closely spaced, rigid, weightless rods along the generators. However, if an inextensible plate, subjected to given loads, had such rods attached in a different alignment, two features are apparent. First, the resultant strain energy would be solely due to bending about these new generators and, second, this strain energy would be *less* than the true value. It follows that the correct distribution of the generators may be determined by maximizing the strain energy associated with an arbitrary distribution of generators, an erroneous distribution being tantamount to the introduction of constraints.

9.3.2 Determination of generators

Consider a thin cantilever plate of variable rigidity under an arbitrary normal loading, such as depicted in Fig. 9.14.

The plate deforms into a developable surface whose generators are identified by an equation of the form

$$\alpha = \alpha(X) \tag{9.55}$$

where α is the orientation of a generator that cuts the x-axis at a point X. Let two typical adjacent generators making angles α and $(\alpha + d\alpha)$ with

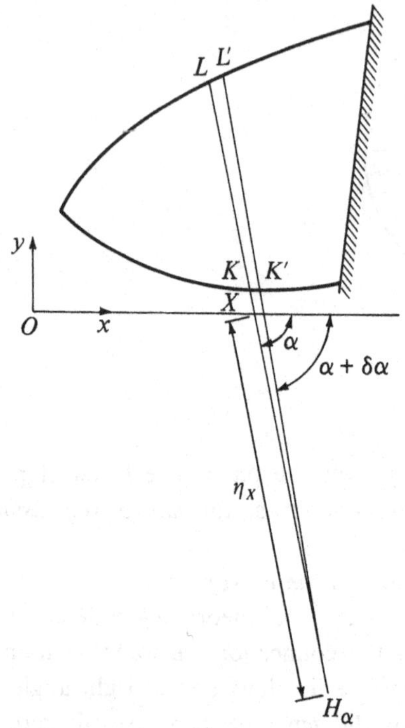

Fig. 9.14

the x-axis intersect on the edge of regression at the point H_α. The elemental slice $KLL'K'$ bounded by these generators undergoes conical bending with the cone apex at H_α. Now let $\boldsymbol{n}$ be directed normal to a generator and in the plane of the plate. The curvature κ_n is the non-zero principle curvature at a point on a generator and hence

$$\kappa_n = c_\alpha/\eta, \text{ say,} \tag{9.56}$$

where c_α, to be determined later from considerations of equilibrium, is a function of the generator angle α, and η is the distance along the α-generator from the point H_α. The moment per unit length about a generator is thus given by

$$M_n = D(\alpha, \eta)c_\alpha/\eta, \tag{9.57}$$

where D is the rigidity and the coordinates α, η define a point in the plate. In this connection, we note that the value of η at the x-axis, η_X say, is given by

$$\eta_X = \pm\frac{\sin\alpha}{\alpha'} \tag{9.58}$$

where a prime denotes differentiation with respect to X, and the sign depends on whether H_α is below or above the x-axis.

Equilibrium

An equilibrium condition is now formed by taking moments about a generator, for then there is no contribution from the unknown middle-surface forces. The moment over an elemental distance $\delta\eta$ is $M_n\delta\eta$, and the total moment about the α-generator, $\mathcal{M}_\alpha$ say, is therefore given by

$$\mathcal{M}_\alpha = \int_{\eta_1}^{\eta_2} M_n \,\mathrm{d}\eta \tag{9.59}$$

where η_1, η_2 are the values of η at the plate boundary.

For a given function $\alpha(X)$ the moment $\mathcal{M}_\alpha$ may be expressed in terms of the loads applied to the plate, and hence (9.57), (9.59) yield the following expression for c_α:

$$c_\alpha = \frac{\mathcal{M}_\alpha}{\displaystyle\int_{\eta_1}^{\eta_2} \frac{D(\alpha,\eta)}{\eta}\,\mathrm{d}\eta}. \tag{9.60}$$

Strain energy

In determining the strain energy in the plate it is convenient to integrate first over an elemental sector bounded by generators, and then to integrate with respect to α. Thus

$$U = \frac{1}{2}\int_{\eta_1}^{\eta_2}\int M_n \kappa_n \eta \,\mathrm{d}\eta\,\mathrm{d}\alpha,$$

$$= \frac{1}{2}\int \mathcal{M}_\alpha c_\alpha \,\mathrm{d}\alpha, \text{ by virtue of (9.56)–(9.60),}$$

$$= \frac{1}{2}\int F\,\mathrm{d}X \quad \text{(say) where } F = \frac{\alpha' \mathcal{M}_\alpha^2}{\displaystyle\int_{\eta_1}^{\eta_2} \frac{D(\alpha,\eta)}{\eta}\,\mathrm{d}\eta}. \tag{9.61}$$

From (9.58) and other geometrical considerations, the function F above is expressible in terms of X, α, α' and the applied loading and plate rigidity.

Determination of $\alpha(X)$

The function $\alpha(X)$ is to be determined from the condition that the strain energy U is a maximum. It follows from the calculus of variations that

$$\alpha'' F_{\alpha'\alpha'} + \alpha' F_{\alpha\alpha'} + F_{X\alpha'} - F_\alpha = 0 \tag{9.62}$$

in which F_α, for example, stands for $\partial F/\partial\alpha$ where F is formally regarded as a function of independent variables X, α, α'. Equation (9.62) is a non-linear differential equation of the second order and, with the boundary conditions, it determines the function $\alpha(X)$. Some simplification of this equation is possible if the function F does not contain X explicitly; for this important class of problems the equation may be integrated once to give

$$F - \alpha' F_{\alpha'} = \text{constant}. \tag{9.63}$$

The solution of (9.62) and (9.63) for a number of examples is given shortly, but first we derive a more physical interpretation of the term c_α which is then used to derive an expression for the plate deflexion.

The term c_α

The form of the second expression for U in (9.61) suggests the following interpretation of the term c_α. We first introduce Ω_α the rotation of the plate about the α-generator, so that

$$\Omega_\alpha = -\frac{\partial w}{\partial \boldsymbol{n}}. \tag{9.64}$$

Now

$$\begin{aligned} \kappa_n &= -\frac{\partial^2 w}{\partial \boldsymbol{n}^2} \\ &= \frac{\partial \Omega_\alpha}{\partial \boldsymbol{n}}, \text{ by virtue of (9.64)}, \end{aligned} \tag{9.65}$$

and

$$\delta \boldsymbol{n} = \eta \delta \alpha, \tag{9.66}$$

so that

$$\kappa_n = \frac{1}{\eta} \frac{\mathrm{d}\Omega_\alpha}{\mathrm{d}\alpha}, \tag{9.67}$$

and a comparison with (9.56) thus shows that

$$c_\alpha = \frac{\mathrm{d}\Omega_\alpha}{\mathrm{d}\alpha}. \tag{9.68}$$

Finally, we note that the energy stored in the plate is necessarily equal to the work done by the applied loads and hence

$$U = \frac{1}{2} \int \mathscr{M}_\alpha \, \mathrm{d}\Omega_\alpha, \tag{9.69}$$

which can, of course, be cast into the form of (9.61).

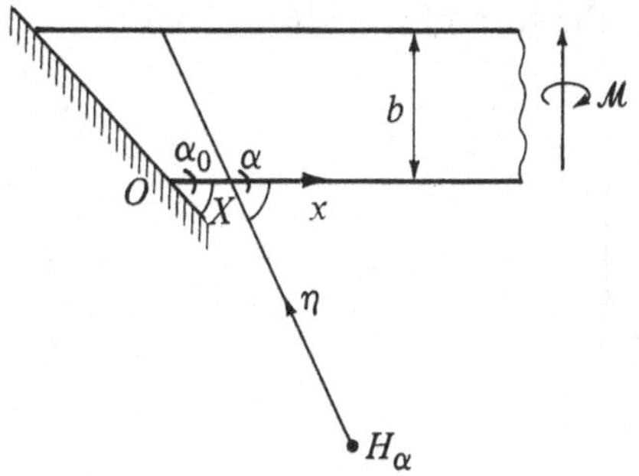

Fig. 9.15

The plate deflexion

Consider a point P at α_P, η_P, and let p_α be the perpendicular from P to an α-generator. The relative rotation $\delta\Omega_\alpha$ between adjacent generators at α, $(\alpha+\delta\alpha)$ is given by

$$\delta\Omega_\alpha = \frac{d\Omega_\alpha}{d\alpha}\delta\alpha$$
$$= c_\alpha \delta\alpha. \tag{9.70}$$

The deflexion of the point P due to this elemental rotation is therefore given by

$$\delta w_P = -p_\alpha c_\alpha \delta\alpha. \tag{9.71}$$

The deflexion of the point P relative to a boundary generator α_B, say, where the plate is clamped is thus given by

$$w_P = -\int_{\alpha_P}^{\alpha_B} p_\alpha c_\alpha \, d\alpha$$
$$= -\int_{X_P}^{X_B} \alpha' p_\alpha c_\alpha \, dX, \quad \text{where } X_P = X(\alpha_P), \ldots \tag{9.72}$$

Note that if the loading consists of a single load $\mathcal{L}$, say, at the point P (positive if in the direction of positive w)

$$\mathcal{M}_\alpha = -\mathcal{L} p_\alpha, \tag{9.73}$$

and (9.61) and (9.72) yield

$$w_P = \frac{2U}{\mathcal{L}}, \text{ as expected.} \tag{9.74}$$

9.3.3 *Swept plate of uniform thickness*

Consider a long strip clamped at one end at an angle α_0, and subjected to a pure moment $\mathcal{M}$ at the far end, as shown in Fig. 9.15. If the x-axis is located along one edge of the strip of width b we have

$$\left.\begin{aligned} \mathscr{M}_\alpha &= \mathscr{M}\sin\alpha, \\ \eta_1 &= \frac{\sin\alpha}{\alpha'}, \\ \text{and} \qquad \eta_2 &= \eta_1 + b\operatorname{cosec}\alpha, \end{aligned}\right\} \tag{9.75}$$

so that (9.61) yields

$$\left.\begin{aligned} F &= \frac{\mathscr{M}^2\alpha'\sin^2\alpha}{D\ln(1+\mu)}, \\ \text{where} \qquad \mu &= \frac{(\eta_2-\eta_1)}{\eta_1} = b\alpha'\operatorname{cosec}^2\alpha. \end{aligned}\right\} \tag{9.76}$$

The function F does not contain X explicitly, and hence (9.62) may be integrated once to yield (9.63), and this can be expressed in the form

$$\frac{\mu^2\sin^4\alpha}{D(1+\mu)\{\ln(1+\mu)\}^2} = C. \tag{9.77}$$

The constant C is to be determined from the condition that as $X \to \infty$, $\mu \to 0$, and $\alpha \to \frac{1}{2}\pi$, whence

$$C = b^2. \tag{9.78}$$

Equations (9.77), (9.78) yield the relation between α and μ which may be further integrated by numerical or graphical means to give the α, X relation in the form

$$X = b\int_{\alpha_0}^{\alpha} \frac{\operatorname{cosec}^2\alpha}{\mu}\,d\alpha. \tag{9.79}$$

Note that as a consequence of the integral form of (9.79), the (α, X) relationship for a given value of α_0 necessarily embodies the solution for all greater values of α_0. This feature, which is typical of all inextensional solutions, stems from the fact that clamping along any generator does not alter the pattern of generators elsewhere.

Once the (α, X) relationship has been determined, the generators of the deflected plate are known and the bending moments and associated stresses may be found from (9.56), (9.58) and (9.60). The bending stresses associated with M_n vary linearly through the plate thickness, reaching a value on the surface given by

$$\sigma = 6M_n/t^2 \tag{9.80}$$

and this is a principal stress. The other principal stress, in the direction of the generators, is of magnitude $\nu\sigma$ because the generators remain straight and so prevent any anticlastic curvature. The maximum value of σ occurs at the junction of the 'trailing edge' and the clamped edge, where $\alpha = \alpha_0$ and $\eta = \eta_1$, and it can be shown that this peak stress is related to the angle α_0 by the equation

$$\left.\begin{aligned} &\sin\alpha_0 = \left(\frac{2\zeta\ln\zeta}{\zeta^2 - 1}\right)^{1/2} \\ &\text{where } \zeta \text{ is a stress concentration factor defined by} \\ &\zeta = \sigma/(\sigma_{x=\infty}) \\ &\quad = \frac{\sigma b t^2}{6\mathcal{M}}. \end{aligned}\right\} \tag{9.81}$$

The closed-form nature of the above equation stems from the fact that the stresses, for a given value of α, depend on the value of η – and hence α' – rather than X. They may thus be derived directly from (9.77) rather than (9.79).

9.3.4 Swept plate of variable rigidity

The case in which the rigidity of the strip varies across the chord may also be readily solved. Consider, for example, a rigidity that varies parabolically across the chord according to the relation

$$D = 4D_0\left\{\frac{y}{b}\left(1 - \frac{y}{b}\right)\right\}. \tag{9.82}$$

Such a variation may be expressed in terms of η by virtue of the equation

$$\frac{y}{b} = \frac{\eta - \eta_1}{\eta_2 - \eta_1} \tag{9.83}$$

and the denominator in the expression for F is now given by

$$\int_{\eta_1}^{\eta_2} \frac{D}{\eta}\,\mathrm{d}\eta = \frac{2D_0}{(\eta_2 - \eta_1)^2}\{\eta_2^2 - \eta_1^2 - 2\eta_1\eta_2\ln(\eta_2/\eta_1)\}. \tag{9.84}$$

Equations (9.61), (9.75), (9.63) and (9.84) suffice to determine an equation connecting α, α', which in turn may be integrated to obtain the (α, X) relation.

9.3.5 Swept plate under arbitrary moment

A similar analysis is applicable to a swept plate of variable rigidity subjected to a moment about an axis making an angle β with the x-axis. The only

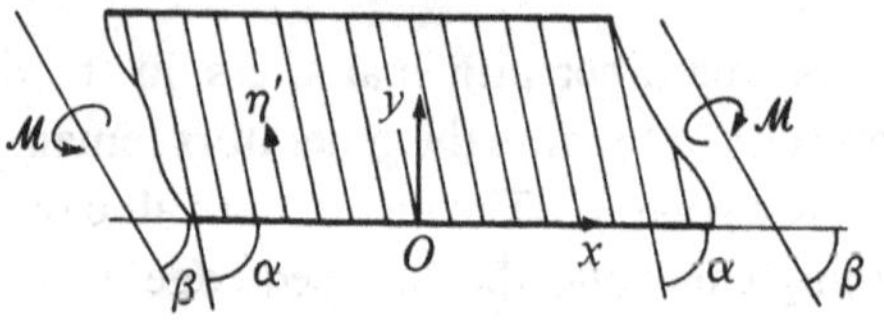

Fig. 9.16

difference in the expression for F lies in the term $\mathcal{M}_\alpha$ which is now given by

$$\mathcal{M}_\alpha = \mathcal{M} \cos(\beta - \alpha). \tag{9.85}$$

As x tends to infinity, the generators become sensibly parallel, but their orientation can scarcely be written down by inspection, as in the case for which $\beta = \frac{1}{2}\pi$. This 'steady state' orientation of parallel generators will now be considered.

9.3.6 *Steady-state deformation of long strip under arbitrary moment* (Fig. 9.16)

When the generators are parallel the analysis of Section 9.3.2 is not immediately applicable because η and c_α become infinite. We may, however, write

$$\frac{M_n}{D(\eta)} = \kappa, \text{ a constant.} \tag{9.86}$$

Now from (9.59), (9.85) and (9.86)

$$\mathcal{M}_\alpha = \mathcal{M} \cos(\beta - \alpha)$$

$$= \int_0^{b \operatorname{cosec} \alpha} M_n \, d\eta', \text{ where } \eta' \text{ is measured from the } x\text{-axis,}$$

$$= \kappa \operatorname{cosec} \alpha \int_0^b D \, dy. \tag{9.87}$$

so that κ is known in terms of the applied loading.

The strain energy of the strip is now given by

$$U = \frac{1}{2} \int_0^l \int_0^b \kappa^2 D \, dx \, dy$$

$$= \frac{1}{2} \int_0^l dx \left(\frac{\mathcal{M}^2 \cos^2(\beta - \alpha) \sin^2 \alpha}{\int_0^b D \, dy} \right), \tag{9.88}$$

and the maximum value of U occurs when

$$\alpha = \tfrac{1}{4}\pi + \tfrac{1}{2}\beta, \tag{9.89}$$

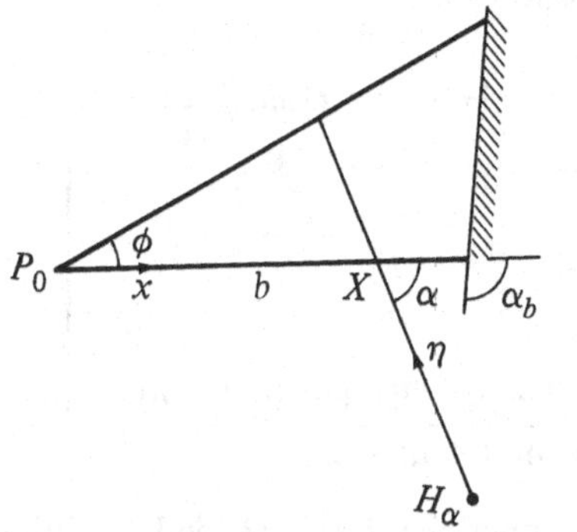

Fig. 9.17

a result which is independent of the chordwise variation of rigidity. Equation (9.89) is also in agreement with the known asymptotic behaviour of the strip of lenticular section considered in Section 7.4.1.

9.3.7 *Triangular cantilever plate under tip load*

The plate with a typical generator is shown in Fig. 9.17. From geometrical considerations it follows that

$$\left.\begin{aligned} \eta_1 &= \frac{\sin\alpha}{\alpha'}, \\ \eta_2 &= \eta_1 + \frac{X\sin\phi}{\sin(\alpha+\phi)}, \end{aligned}\right\} \tag{9.90}$$

and if $\mathscr{L}$ is the tip load

$$\mathscr{M}_\alpha = -\mathscr{L}X\sin\alpha. \tag{9.91}$$

For a plate of constant thickness we therefore have

$$\left.\begin{aligned} F &= \frac{\mathscr{L}^2X^2\alpha'\sin^2\alpha}{D\ln(1+\mu)}, \\ &\text{where} \\ \mu &= \frac{\eta_2-\eta_1}{\eta_1} = \frac{X\alpha'\sin\phi}{\sin\alpha\sin(\alpha+\phi)}. \end{aligned}\right\} \tag{9.92}$$

The function F thus depends on X, α and α', and (9.62) does not admit of an immediate integration. However, F is homogeneous in X, and hence (9.62) is homogeneous and can therefore be expressed as a first-order differential equation in α and μ (say); in physical terms homogeneity means that the pattern of generators is independent of the size of the plate. Thus, after some manipulation, it is found that

$$\left.\begin{aligned}
&\sin\alpha\sin(\alpha+\phi)\frac{d\mu}{d\alpha}+\frac{2J}{\mu}\sin(2\alpha+\phi)+Q\sin\phi=0,\\
&\text{where}\\
&J=\frac{\mu^2(1+\mu)\ln(1+\mu)}{(2+\mu)\ln(1+\mu)-2\mu},\\
&Q=\frac{(1+\mu)\ln(1+\mu)\{\mu(2+\mu)-2(1+\mu)\ln(1+\mu)\}}{\mu\{(2+\mu)\ln(1+\mu)-2\mu\}},
\end{aligned}\right\} \tag{9.93}$$

and we note that as $\mu\to 0$, the functions J and Q tend to finite values, namely 6 and 2, respectively.

The integration of (9.93) may be done numerically using a step-by-step process involving the boundary conditions at the tip, discussed shortly. Once this primary integration is completed, the further integration to give the X, α relation is straightforward because, from (9.92),

$$\frac{1}{X}\frac{dX}{d\alpha}=\frac{\sin\phi}{\mu\sin\alpha\sin(\alpha+\phi)}, \tag{9.94}$$

which may be integrated to give

$$X=C\exp\int_{\alpha_0}^{\alpha}\frac{\sin\phi\, d\alpha}{\mu\sin\alpha\sin(\alpha+\phi)}, \tag{9.95}$$

where C is chosen so that $\alpha=\alpha_b$ at $X=b$, its value at the clamped boundary.

The tip boundary condition

The generator angle α_0 at the tip depends only on the tip angle ϕ; it is independent of the orientation of the root fixing. This feature is a specialization of Saint Venant's principle, and it is related to the fact that clamping of a plate *at a point* is, in effect, indistinguishable from other methods of support. Thus it is not possible to impose an arbitrary generator angle at the tip, and this is reflected in the form of (9.93) which, as $\mu\to 0$, yields an infinite value for $d\mu/d\alpha$ except when α_0 assumes its 'natural' value given by

$$\alpha_0=\tfrac{1}{2}(\pi-\phi). \tag{9.96}$$

Thus the generator at the tip is normal to the bisector of the tip angle.

9.3.8 Triangular cantilever plate under uniformly distributed load

For this case η_1 and η_2 are again given by (9.90), while for a unit load/unit area

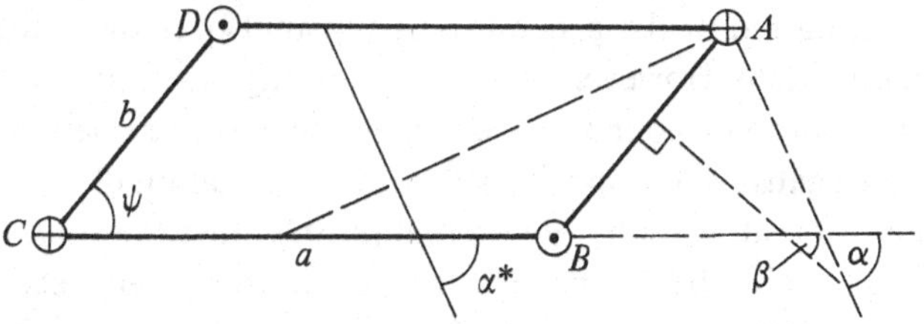

Fig. 9.18

$$\mathscr{M}_\alpha = -\frac{X^3 \sin\phi \sin^2\alpha}{6\sin(\alpha+\phi)}, \tag{9.97}$$

so that

$$F = \frac{X^6 \alpha' \sin^2\phi \sin^2\alpha}{36D \sin^2(\alpha+\phi)\ln(1+\mu)}. \tag{9.98}$$

Thus F is again homogeneous in X and, corresponding to (9.93), we now find

$$\sin\alpha \sin(\alpha+\phi)\frac{\mathrm{d}\mu}{\mathrm{d}\alpha} + \frac{2J}{\mu}\sin(2\alpha+\phi) + 3Q\sin\phi = 0, \tag{9.99}$$

and the subsequent integration follows a similar path to that in Section 9.3.7.

9.3.9 *Parallelogram plate loaded at the corners*

In Section 9.3.7 it was shown that the generators near the loaded tip of a triangular plate are normal to the bisector of the tip angle. Thus, near the corners A, C of the plate shown in Fig. 9.18,

$$\alpha = \tfrac{1}{2}(\pi - \psi). \tag{9.100}$$

Further, from Section 9.3.6 the orientation of the generators in a long strip under arbitrary moment is given by (9.89). Now for the corner-loaded parallelogram plate, the loading – away from the vicinity of the acute corners A, C – is equivalent to a moment Pb about a normal to the sides AB, CD. Thus, referring to Fig. 9.18 we have

$$\beta = \tfrac{1}{2}\pi - \psi, \tag{9.101}$$

and hence, from (9.89),

$$\begin{aligned}\alpha^* &= \tfrac{1}{4}\pi + \tfrac{1}{2}(\tfrac{1}{2}\pi - \psi)\\ &= \tfrac{1}{2}(\pi - \psi),\end{aligned} \tag{9.102}$$

which is the same as that near the corners A and C. We therefore deduce that the orientation of the generators is constant over the whole of the parallelogram plate. However, instead of focusing attention on the acute corners A, C we could equally have chosen the obtuse corners B, D. This yields an alternative solution with a constant orientation of generators at right angles to that given by (9.100) and (9.102). Both the resulting plate deflexions are stable, but in the former case the strain energy exceeds that in the latter by the factor $\cot^4 \frac{1}{2}\psi$. The former configuration will thus be preferred unless some external agency forces the plate into the other configuration. Examples demonstrating these two modes are given in Section 9.4.2 where they are shown to be intimately related to tension field modes in a parallelogram membrane. For the corner-loaded square or rectangular plate, either mode is equally likely. If we follow the deformation of a square plate, say, as the corner loads $\pm P$ increase, an exact large-deflexion solution would predict a gradual change in the small-deflexion shape as P increases, resulting in a stiffening of the plate as middle-surface forces are introduced. During this phase the anti-symmetric nature of the small-deflexion mode is maintained, in that the deflexion along one diagonal is equal in magnitude but opposite in sign to that along the other diagonal. However, at a critical value P^*, say, a bifurcation occurs. As P increases beyond P^*, the mode shape gradually changes so that the deflexion along one diagonal increases while the other decreases, thus approaching the inextensional mode shape. Experimental results relating the load P to the corner deflexion difference w, where

$$w = \tfrac{1}{2}(w_A + w_C - w_B - w_D),$$

are shown in Fig. 9.19 for three different square plates in which $t/a = 0.004$, 0.007 and 0.011, respectively. In accordance with the non-dimensional terms introduced in Section 7.1.2, the ordinate is Pa^2/Dt and the abscissa is w/t, and it is seen that this presentation condenses the experimental results to a common curve. Unfortunately, there is no known exact large-deflexion solution for comparison, but it is clear that

$$\frac{P^* a^2}{Dt} \approx 20,$$

and that when $P > P^*$ the *stiffness* of the plate is given quite accurately by inextensional theory, although the deflexion w exceeds the inextensional value by an approximately constant amount, namely $3.2t$. Of course, the corner-loaded square plate is an extreme case in that it maximizes the difference between small-deflexion theory and inextensional theory. In this respect it is similar to the torsion of the strip of lenticular parabolic section considered in Section 7.4.1. There, however, the torque-twist relation is given *exactly* by inextensional theory when the torque exceeds the

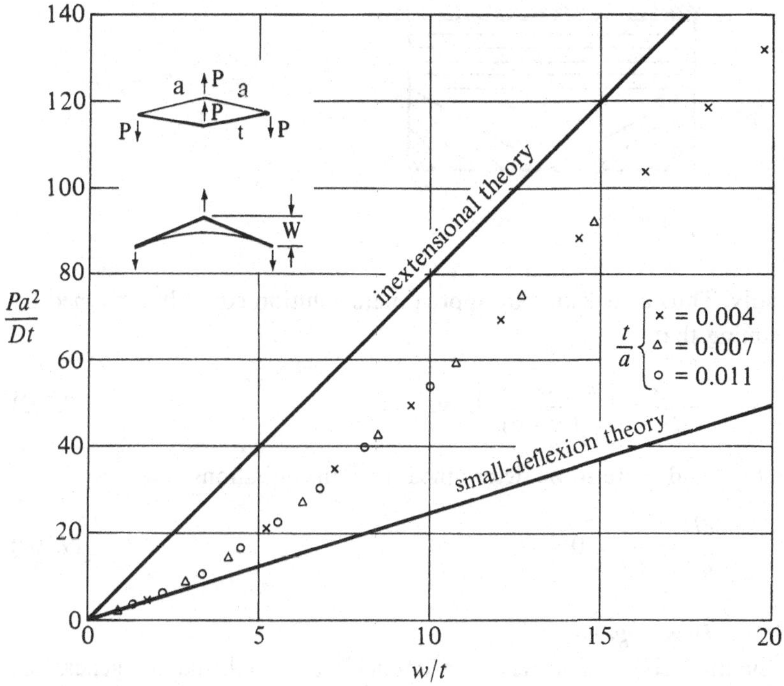

Fig. 9.19

bifurcational value. For most cantilever plates the difference between inextensional theory and small-deflexion theory is much less, while for the strip under pure moment – such as considered in Section 7.4.1 – the stiffness according to inextensional theory exceeds that according to small-deflexion theory by the relatively insignificant factor $(1-\nu^2)^{-1}$.

9.3.10 Approximate inextensional solutions

An approximate solution can be obtained by assuming a suitable relation between X and α which contains one or more arbitrary parameters which are to be determined from the condition that the strain energy is to be a maximum. For example, in the triangular cantilever plates considered previously, the solution for any given root angle α_b is embodied in the solution for

$$\alpha_B, \text{ say} = \pi - \phi, \quad \text{at } X = B,$$

which corresponds to the limiting case in which the root fixing is parallel to the leading edge. The limiting distance B could, of course, be determined from the previous analysis, but even without foreknowledge of the value of B we know that for small values of X, $\alpha \to \frac{1}{2}(\pi - \phi)$ and as $X \to B$, $\alpha \to (\pi - \phi)$. Furthermore, as X approaches B the angle α varies very

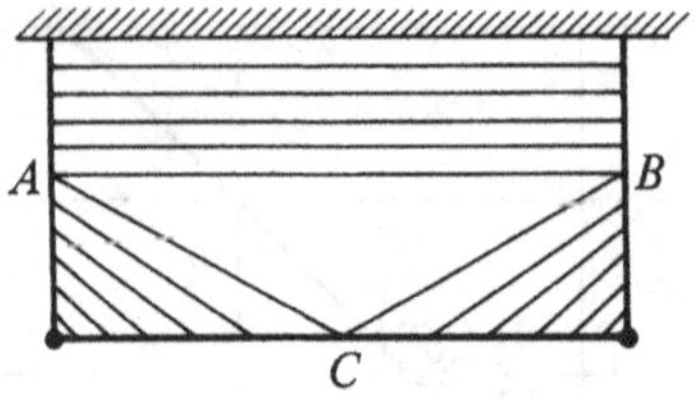

Fig. 9.20

rapidly. Thus a realistic but approximate solution could be obtained by assuming that

$$\frac{B-X}{B}=\left(\frac{\pi-\phi-\alpha}{\frac{1}{2}(\pi-\phi)}\right)^n, \text{ say,} \tag{9.103}$$

where B and n are to be determined from the equations

$$\frac{\partial U}{\partial B}=\frac{\partial U}{\partial n}=0. \tag{9.104}$$

9.3.11 Dead regions

In the analysis so far it has been implicitly assumed that the generators cover the entire surface of the plate and, indeed, this is generally the case. However, there are circumstances in which, according to inextensional theory, a clearly defined region of the plate simply undergoes a rigid body movement. Such regions are called *dead regions*; they are typified by the triangle ABC in Fig. 9.20 which shows the pattern of generators over a rectangular cantilever plate carrying equal concentrated loads at the far corners. In this example the three sides of the triangle AB, BC, CA form the bounding generators of three distinct generator fields; the only developable surface for the triangular region ABC is therefore the plane in which these bounding generators lie. Inside the region ABC there are thus no bending stresses, and the moment about a bounding generator, which is carried by bending stresses on the far side of the bounding generator, must necessarily be carried on the near side by middle-surface stresses. Furthermore, a section across the plate on the near side of a bounding generator is undistorted except in the immediate vicinity of the corner points A, B, C and it follows that the middle-surface stresses are infinite at these corner points.

In practice, the position is not so severe, because a certain amount of stretching of the middle surface, and consequent bending, takes place in the dead regions, thus ironing out the theoretical peaks of stress; but inextensional solutions which include dead regions will overestimate the stiffness of the plate unless allowance is made for their flexibility.

9.4 The analogy between tension field theory and inextensional theory

An examination of the analyses for tension field theory and inextensional theory shows that various terms are analogous, as shown in the following table:

Tension field theory	Inextensional theory
Et	$1/D$
Δ_α	$\mathscr{M}_\alpha$
ε_η	M_n
U	U
F	F
(α, X)	(α, X)

The pattern of tension rays/generators in membranes and plates of the same shape will thus be identical, provided the boundary conditions are also analogous. In this connection we note that in tension field theory a free edge coincides with a tension ray, whereas in inextensional theory a supported edge coincides with a generator. Other analogous boundary conditions stem from the Δ_α, $\mathscr{M}_\alpha$ analogy, and they are summarized in the following table.

First we note that because Δ_α depends only on the displacements *at the boundary* a corollary to this is that the analogy is restricted to cases in

Tension field theory	Inextensional theory
Free edge	Supported edge
Supported edge	Free edge
In a membrane with opposing supported boundaries undergoing a relative rigid-body displacement u, parallel to the x-axis, and v, parallel to the y-axis: u, v.	In a plate with opposing free boundaries, carrying a torque T about the x-axis and a moment M about the y-axis: T, M.
Membranes in which at a boundary corner the adjacent supported edges undergo a relative rotation ω in the plane of the membrane: ω.	Plates in which a boundary corner with adjacent free edges carries a concentrated load P normal to the plane of the plate: P.

inextensional theory in which the loads and moments are applied *at the boundary*.

Some care must be exercised in interpreting this analogy. First, a given inextensional solution does not necessarily imply an analogous tension field solution unless it can be verified that the inequalities (9.21), (9.22) are satisfied. Second, we note that in inextensional theory the α, X relation is unaffected by a reversal of the applied loads, whereas in tension field theory a reversal of the boundary displacements will cause a tension field (if at all) with quite different characteristics. This naturally raises the question in inextensional theory as to the analogy of this further tension field. The answer is simply that *both* analogous systems of generators correspond to *local* stable states. In general, of course, these states contain different strain energies and there will be a preference for the state which contains the greater energy. A simple case in which there is no preference is that of the corner-loaded square plate with free edges; a deflexion pattern with generators parallel to either diagonal is equally probable, while the two tension field analogies correspond to the positive and negative shear distortion of a square membrane. We will shortly consider some other specific cases of the analogy, but first we note that for the practically important case in which the thickness t of the membrane/plate is constant, the analogy between ε_η and M_n means that the corresponding stresses σ_η and σ_n vary in a similar way over the membrane/plate. In particular, distributions of non-dimensional stress concentration factors are equally applicable to the membrane or plate.

9.4.1 *Examples of the analogy*

It may be verified that the shearing of a semi-infinite membrane strip, considered in Section 9.2.4, is the analogue of the swept plate subjected to a torque at the far end. Further, the pattern of tension rays and stress concentration factors shown in Fig. 9.6 is equally applicable to the pattern of generators and stresses in the analogous plate. In fact, Fig. 9.6 embodies the solution for *both* stable inextensional states. Thus, if the sweep-back angle is 15°, say, the *preferred* inextensional solution is represented by that part of Fig. 9.6 to the right of the ray at $\alpha = 75°$. The other stable state is represented by the *whole* of Fig. 9.6, in which the shaded triangle to the left of the ray at $\alpha = 90°$ is a *dead region.*

By the same token, the triangular cantilever plate under tip load is analogous to a triangular membrane bounded by a free edge and two supported edges that undergo a relative rotation about their common vertex. Likewise, the torsion of the slit annular membrane, considered in Section 9.2.6, is analogous to an axially loaded flat annular spring. Finally, the corner-loaded parallelogram plate, considered in Section 9.3.9, is

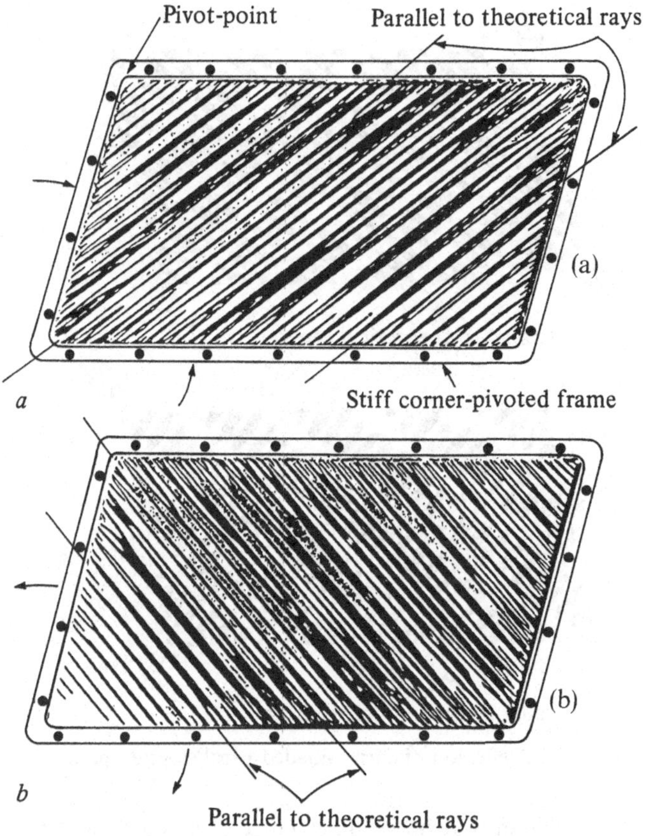

Fig. 9.21 Tension rays in parallelogram membrane.

analogous to a parallelogram membrane whose sides undergo rigid-body rotations about the corners. Because of the simplicity of the theoretical solutions, this particular analogy has been tested to provide a dual experimental check on both theories.

9.4.2 Experimental results for the parallelogram membrane/plate

Experiments to verify the above theoretical results, and hence the analogy, have been made on a parallelogram membrane and plate of identical shape. The sides of each measured 25.4 cm by 14.5 cm with $\psi = 75°$. The membrane was of aluminized polyester 0.0015 cm thick and the plate was of silver-plated spring steel 0.02 cm thick; both had good light-reflecting surfaces. The edges of the membrane were clamped to a stiff brass frame with pivots at the corners of the membrane; the membrane dimensions cited exclude the dimension necessary for edge clamping. Distortion of the membrane was effected by screwing together a pair of opposite corners.

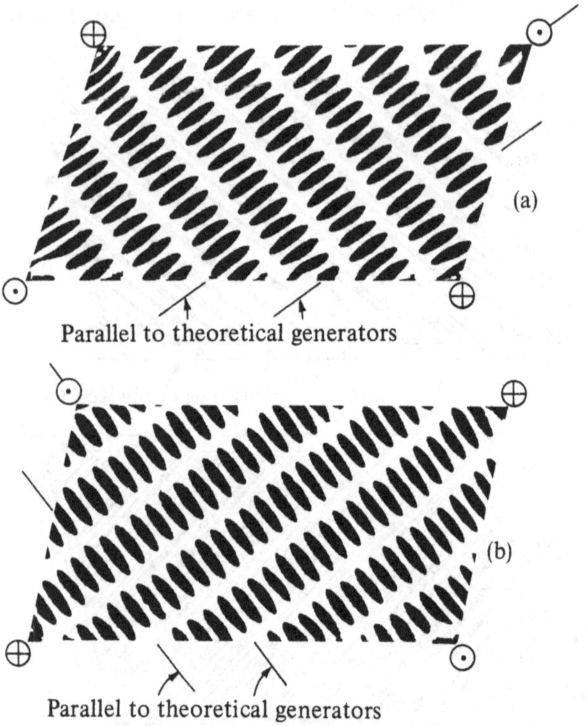

Fig. 9.22 Deflexion of corner-loaded parallelogram plate.

Fig. 9.21*a* and *b* shows the wrinkles corresponding to a contraction of the shorter and longer diagonals, respectively. These show good agreement with theory.

The corner-loaded plate was supported in a vertical plane and positioned so that the theoretical generators would remain vertical after load. The plate was then photographed through a (slightly) curved white screen on which was painted a chessboard pattern of black circles, a central circle being cut out to provide an aperture for the camera. According to inextensional theory, the plate deforms into a purely cylindrical form so that the reflection in the plate surface of the patterned screen should consist of a regular array of ellipses whose major axes coincide with the generators. Fig. 9.22*a* and *b* show good agreement with theory over most of the plate but with an expected deviation towards the edges due to the 'boundary layer' effect discussed in Section 7.4.1.

References

Föppl, A. *Vorlesungen über Technische Mechanik*, **5**, p. 132 (1907).
Iai, T. *J. Soc. Aero. Sci. Nippon*, **10**, p. 96 (1943). [In Japanese.]
Jahsman, W. E., Field, F. A., and Holmes, A. M. C. Finite deformations in a

prestressed, centrally loaded circular elastic membrane. *Proc. 4th Nat. Cong. Appl. Mechs.*, **1**, pp. 585–94 (1962).

Kondo, K. *J. Soc. Aero. Sci. Nippon*, **5**, p. 41 (1938). [In Japanese.]

Kondo, K., Iai, T., Moriguti, S., and Murasaki, T. *Memoirs of the unifying study of the basic problems in engineering sciences by means of geometry*, I. Gakujutsu Bunken Fukyu-Kai, Tokyo, 1955.

Mansfield, E. H. The inextensional theory for thin flat plates. *Quart. J. Mech. Appl. Maths.*, **8**, pp. 338–52 (1955).

——. Tension field theory, a new approach which shows its duality with inextensional theory. *Proc. XII Int. Cong. Appl. Mech.*, pp. 305–20 (1968).

Mansfield, E. H., and Kleeman, P. W. A large-deflexion theory for thin plates. *Aircraft Engineering*, **27**, pp. 102–8 (1955a).

——. Stress analysis of triangular cantilever plates. *Aircraft Engineering*, **27**, pp. 287–91 (1955b).

Schwerin, E. Uber Spannungen und Formanderungen kreisringformiger Membranen. *Zf. Tech. Physik.*, **12**, pp. 651–9 (1929).

Stein, M., and Hedgepeth, T. M. Analysis of partly wrinkled membranes. *NASA Tech. Note* D-813 (July 1961).

Wagner, H. Ebene Blechwandträger mit schr dünnem Stegblech. *Z. Flugtech. u. Motorluftschs.*, **20**, Nos. 8, 9, 10, 11, 12 (1929).

Additional references

Ashwell, D. G. The equilibrium equations of the inextensional theory for thin flat plates. *Quart. J. Mech. Appl. Maths.*, **10**, pp. 169–82 (1957).

Fung, Y. C., and Wittrick, W. H. A boundary layer phenomenon in the large deflexion of thin plates. *Quart. J. Mech. Appl. Maths.*, **8**, pp. 191–210 (1955).

Kuhn, P. Investigations on the incompletely developed diagonal tension field. *N.A.C.A. Rep.* No. 697 (1940).

Mansfield, E. H. Load transfer via a wrinkled membrane. *Proc. R. Soc. Lond.* Ser. A, **316**, pp. 269–89 (1970).

——. Analysis of wrinkled membranes with anisotropic and nonlinear elastic properties. *Proc. R. Soc. Lond.* Ser. A, **353**, pp. 475–98 (1977).

Riparbelli, C. Non extensional deformation modes of thin plates. *IX Congrès International de Méchanique Apliquée*, Actes, **6**, pp. 442–7 (1957).

Simmonds, J. G. Closed-form, axisymmetric solution of the von Kármán plate equations for Poisson's ratio one-third. *J. Appl. Mech.*, **50**, pp. 897–8 (1983).

AUTHOR INDEX

SUBJECT INDEX